WETLANDS
IN DANGER

WETLANDS
IN DANGER

General Editor Patrick Dugan
Introduction by David Bellamy

Published in association with
IUCN – The World Conservation Union

Mitchell Beazley

Contributors

Eastern Canada and Greenland David Boertmann, Steffen Brogger Jensen, Clayton Rubec

Western Canada and Alaska Tom Dahl, Clayton Rubec, Jim Thorsell

The United States – The Lower "Forty-Eight" Tom Dahl, Joe Larson, Dan Scheidt,

Mexico, Central America and the Caribbean Alejandro Yanez Arancibia, Peter Bacon, Monica Herzig, Enrique Lahmann

Northern South America and the Amazon Basin Antonio Diegues, Stefan Gorzula, Francisco Rilla

Southern South America Argentino Boneto, Pablo Canevari, Maria Marconi, Victor Pullido, Francisco Rilla, Hernan Verascheure, Yerko Vilina

Northern Europe David Boertmann, Steinar Eldoy, Jens Enemark, Palle Uhd Jepsen, Esko Joutsamo, Torsten Larsson, Karsten Laussen, Hans Meltofte, Ole Thorup

West and Central Europe Andrew Craven, Nic Davidson, P. Gatescu, Liz Hopkins, Zbig Karpowicz, Edward Maltby, Francois Sarano, A. Vadineanu, Jurgen Voltz, Edith Wenger

The Mediterranean Basin George Catsadorakis, Alain Crivelli, A. Gerakis, Alain Johnson, Thymio Papayannis, Jamie Skinner, Nergis Yazgan

The Middle East Andrew Price, Derek Scott

East Africa and the Nile Basin Geoff Howard, Paul Mafabi, Steven Njuguna, M. A. Zahran

West and Central Africa Pierre Campredon, Jean-Yves Pirot, Ibrahima Thiaw

Southern Africa Geoff Howard, Tabeth Matiza, Rob Simmons

Northern Asia Genady Golubev, Vitaly G. Krivenko, Mike Smart, Vadim G. Vinogradov

Central and South Asia Zakir Hussain, Peter-John Meynell, Sam Samarakoon

East Asia Derek Scott, Satoshi Kobayashi

Southeast Asia Zakir Hussain, Duncan Parish, Marcel Silvius

Australia Jim Davie, Max Finlayson, Jim Puckeridge

New Zealand and the Pacific Department of Conservation (New Zealand), Derek Scott

Senior Executive Editor
Robin Rees

Senior Editor
Steve Luck

Editors
William Hemsley
Clint Twist

Senior Designer
Jean Jottrand

Picture Research
Kathy Lockley

Production
Katy Sawyer

Maps
World Conservation
  Monitoring Centre
Cambridge Desktop
  Bureau

Proofreading
Fred Gill

Indexing
Ann Barret

First published in Great Britain in 1993
by Mitchell Beazley
an imprint of Reed Consumer Books Limited
Michelin House, 81 Fulham Road, London SW3 6RB
and Auckland, Melbourne, Singapore and Toronto

Copyright © Reed International Books Limited, 1993
All rights reserved

A CIP catalogue record for this book is available at the British Library.

ISBN 1 85732 166 9

Typeset in Bembo and Trade Gothic
Reproduction by Mandarin Offset, Singapore
Produced by Mandarin Offset
Printed and bound in Singapore

Measurements
Both metric and imperial measurements are given throughout
Billions The usage of the word billion varies between countries, but in all cases here, it refers to thousand millions (1,000 million = 1 billion)
Tonnes/tons All measurements are given in tonnes (for precise conversion, 1 tonne = 0.98 tons)

Jacket pictures
Front An Australian jacana (*Irediparra gallinacea*) in Kakadu National Park in the Northern Territory, Australia. (Planet Earth Pictures)
Back top Fishermen off the coast of Michoacan, Mexico. (ZEFA)
Back bottom Cypress pines of Big Cypress Swamp in the Everglades, Florida, USA. (Planet Earth Pictures)

22.5.95.

Contents

Introduction

The twenty-two species of the genus *Hydrostachys* give us an anchor point at the fast flowing end of the biodiversity of the world's wetlands. Members of a tiny family of flowering plants, they live attached to rocks in waterfalls and torrents. There they enjoy plenty of Sun and oxygen – their main problem is hanging on. At the other end of the line are the *Sphagna*, very special mosses which thrive in the oxygen-deficient, mineral-poor stagnant water of the acid peatlands which they help to create. Between these two extremes is the whole diversity of wetlands – springs, streams, rivers, ponds, lakes, oxbows, swamps, fens, water meadows, estuaries, and intertidal mangrove marsh, even semi-desert landscapes have their *waddis* (temporary water courses).

At the heart of every living landscape is, or rather was, a complex series of wetlands, habitats for fish amphibians, reptiles, ducks and other waterfowl, kingfishers and herons, otters, beavers, seals, manatees, hippos, dolphins and many other endangered animals. Water lilies, water hyacinth, kariba weed and duckweed vie for surface space, while the undersurface of their leaves are habitats for smaller plants and animals, such as algae, hydras, flatworms and snails. A drop of wetland water viewed under a microscope can give hours of fascination, watching diatoms and volox glide by while water bears and other mini monsters play their voracious part near the base of a complex food chain. A food chain complete with producers, consumers, and decomposers which given half a chance purifies the water again and again as it flows to the sea.

Crustaceans, shrimps, crabs, and crayfish abound in both fresh and salt water, but some strange quirk of evolution has kept the most successful group of all, the insects, from exploiting the sea. Not so the wetlands and one of the signs of a landscape in a healthy state is the abundance of dragon-, damsel-, may-, and alder flies, all of which spend most of their lives as part of the wetland communities.

Preserving the balance

Under natural conditions things appear to go wrong only in those places where organic matter is not broken down but accumulates as peat. A similar process in the past laid down the vast deposits of coal on which 20th-century society has come to depend. As the coal formed, it sequestered carbon dioxide from the atmosphere, producing oxygen into the bargain. Peat-forming wetlands, coral reefs and some marine plankton are today the only things that perform the vital function of keeping the gases of the atmosphere in balance. They are the true "lungs" of the Earth. The world's peatlands alone cover more than 2.3 million square kilometres (900,000 square miles) with 330 billion dry tonnes of organic matter. If burned or drained and so opened up to slow oxidation, they would release 500 billion tonnes of carbon dioxide, almost double the amount present in the atmospheric greenhouse today.

The continued destruction of wetlands by drainage, exploitation and pollution is the worst act of environmental vandalism being committed on a world-wide scale today. Teeth must be put into the Ramsar Convention by the rich, developed countries putting their own wetlands in order. How can we expect desperate, poor developing countries to look after their wetland heritage, when we continue to destroy ours?

It has been calculated that 1 hectare (2.5 acres) of tidal wetland can do the job of US$123,000 worth of state of the art waste-water treatment, and many communities and companies are now recreating wetlands to cleanse their waste. Yes there are lights at the end of the tunnels, but time is running out for so many crucially important sites and for the world at large, for without wetlands the biosphere cannot continue its vital work.

David J. Bellamy
The Conservation Foundation,
London

Right **The Pantanal of Brazil. The seasonally inundated Pantanal is a vast patchwork of palm savannas, rivers, freshwater lakes and marshes. Despite its vast size, half that of France, only two small areas of the Pantanal have acquired a protected status. The Mato Grosso State Government has, however, banned all gold mining in the region, and fish exports have been stopped in an attempt to conserve the dwindling fish reserves.**

Foreword

Wetlands have been treated with such hostility by many human societies over so many years that their conservation seems almost counter-cultural. Grendel, the monster in *Beowulf,* the earliest epic in the English language, "held the moors, the fen and the fastness" and ravaged Seeland from his swampy home.

The drainers of the fens of the Netherlands and eastern England, and later of the floodplains of North America, were hailed as social heroes, making corn grow where only reeds and sedges once prevailed. Straightening and embankment has transformed most of the great rivers of Europe and many of those in North America. The Nile has been checked by the great dam at Aswan, and the Zambezi by Lake Kariba. China is about to dam the Yangtze, while in New Zealand, there is controversy over the fate of the last natural braided river that traverses South Island south of Christchurch. And so on.

Yet wetlands are far from the useless, fever-ridden, desolations that such actions might imply. They are rich in fish, birds, crocodiles and other valuable creatures. They support herbs with high medicinal value. They are the natural habitat of one of the world's principal food grains – rice – most strains of which are cultivated in a modified wetland habitat, often tilled by a wetland mammal, the water buffalo. And estuarine and coastal wetlands – mangrove forests, salt marshes, sea-grass beds and mud flats – have enormous biological productivity and are important as nursery grounds for marine fishes as well as defences of low-lying coasts against raging sea storms.

And wetlands are regulators of water flow. Many a river remains a reliable source of water throughout the year because its flow is impeded by swamplands so that wet season downpours drain away slowly and the water continues to flow during the dry season. Where streams are straightened, and uplands denuded of forests and swamps, flash floods cause chaos after storms, and the accelerated run-off leaves little to rely on in summer drought.

Counting the cost

Today, many societies are learning the foolishness and cost of treating wetlands as a public enemy. Cities that have encroached on floodplains have to bear the tragedy and cost of avoidable floods. Some communities are restoring wetlands as a simple, cheap, effective way of cleaning up sewage effluent. Others are recognizing the enormous value of fresh and brackish-water wetlands as sources of fish and other foods. Vast areas of the Chinese coastland are covered by fishponds vital to local communities.

The aesthetic quality of wetlands is also gaining increasing recognition. Lakes, rivers and estuaries are being defended ever more vigorously against those who would engineer them away. Their value for fishing, boating, as a haunt of wild birds, and as wide-open landscapes full of colour, reflecting the changing patterns of the clouds is appreciated by ever more people. It was the poet Gerard Manley Hopkins who wrote:

"What would the world be, once bereft
Of wet and of wildness? Let them be left!"

As this book shows, wetlands have been drained, shrunken and parcelled more severely than almost any other wild habitat, and humanity has lost in the transformation. We need to plan and manage our remaining wetlands far more wisely than in the uncritical years of the drainer. Some such land may need to be taken for farmland to feed a world where human populations and human hunger are both increasing steadily. But this must be a deliberate process, informed by understanding.

I hope that this book will help create that understanding among thinking people throughout the world.

Martin W. Holdgate
Director General of IUCN

Right **The largest inland delta in the world, the Okavango Delta, lies on the northern edge of the Kalahari Desert in Botswana. The Okavango Delta, about half of which is permanent swamp and the rest seasonal floodplain, is home to some of Africa's rarest species of bird and other wildlife. The importance of the delta as a refuge for animals is reflected by the fact that a large proportion has been made a wildlife preserve, the Moremi Game Reserve.**

Damp morning conditions in central Finland. Droplets of dew have condensed on the spiky sedge grass of a forest bog. Among the trees, where the atmosphere is slightly warmer, the morning mist still clings. Finland has an exceptionally high proportion of wetlands – 11 per cent of the country's surface area is covered by approximately 55,000 lakes; a further 30 per cent of the land consists of marsh and bog.

What Are Wetlands ?

Most of us are familiar with wetlands in some shape or form. The village pond, the trout stream or the local estuary, for example, are just three of the many types of wetland that are widespread throughout temperate regions. But further south, in tropical and sub-tropical regions, there are muddy tidal flats, expansive floodplains and misty swamplands: three very different environments with very different plants and animals, but these are wetlands too.

Wetland Diversity

There are more than 50 definitions of wetlands in use throughout the world. Among these the broadest, and therefore that which is used most widely on an international scale, is provided by the *Ramsar Convention on Wetlands of International Importance, Especially as Waterfowl Habitat*. Ramsar is an Iranian city lying on the shores of the Caspian Sea, and it was here that the Wetland Convention was adopted in 1971. Designed to provide international protection to the widest possible group of wetland ecosystems, the Ramsar Convention defines wetlands as: "areas of marsh, fen, peatland or water, whether natural or artificial, permanent or temporary, with water that is static or flowing, fresh, brackish or salt, including areas of marine water the depth of which at low tide does not exceed six metres [20 feet]."

Estuaries, mangroves and tidal flats

Estuaries form where rivers enter the sea. The daily tidal cycle and the intermediate salinity between salt and fresh water, which are characteristic of these ecosystems, make them difficult places in which to live. However, those species that have adapted successfully thrive in these conditions. Indeed, estuaries and inshore marine waters are among the most naturally fertile habitats in the world.

Estuaries are found in all regions of the world, but their productivity varies with climate, hydrology (the water cycle) and coastal land forms. Many estuaries are associated with important lagoon systems, some of which have been created by the closure of one of the estuaries' outlets to the sea. In temperate regions, intertidal mud and sand flats, salt marshes and scattered, rocky outcrops are common features of estuaries.

In the tropics and sub-tropics, however, mangroves dominate many coastal habitats and are characteristic of most estuaries.

Variously referred to as "coastal woodland", "tidal forest" and "mangrove forest", mangroves comprise very diverse plant communities, whose composition varies greatly from region to region. Even within the same delta, the composition of the mangrove community can vary substantially according to the conditions of salinity, tidal system and substrate (the soil foundation). Approximately

Below Reed boats on Lake Titicaca, the world's highest navigable lake. Situated in the shadow of the towering Cordillera Real mountains on the border between Bolivia and Peru, Titicaca's surface is some 3,810 m (12,500 ft) above sea level. Large stretches of the lake's shoreline consist of marsh and reedbeds. Lake Titicaca, surrounded by high-altitude desert, provides an attractive environment for many species of fish and birds, as well as a population of tens of thousands of people. The area has been inhabited for several thousand years, and the potato is believed to have been first cultivated in terraced fields around Titicaca's shores.

80 species of plant are recognized as being mangroves. They share a variety of adaptations that enable them to grow in the unstable conditions of estuarine habitats in the tropics and sub-tropics.

Although mangroves, mud flats and other coastal wetland habitats are normally most extensive around estuaries, they are also found along areas of open coast. For example, in Mauritania, the Banc d'Arguin, Africa's largest system of tidal flats, receives no significant surface inflow of freshwater. And sandy beaches, which are characteristic of almost every coastal country, support important populations of wildlife, including migratory shorebirds and nesting marine turtles.

Floodplains and deltas
As rivers swell with seasonal rainfall they slowly rise above the river channel and under natural conditions flow out over the neighbouring plain. This pattern of seasonal flooding was once a common feature of most of the world's rivers. Today, however, with the increasingly widespread construction of dams and embankments, the natural patterns of flooding have been severely disrupted in many regions. Nevertheless, the annual cycle of inundation and desiccation of the world's floodplains remains one of the most important forces governing wetland productivity.

In many areas, floodplains are found in coastal lowlands and end in estuarine deltas where they become complex mosaics of marine, brackish and freshwater habitats. Alternatively, some of the world's larger rivers spread out over floodplains far inland, many of them covering vast areas that include grassy marshes, flooded forest, oxbow lakes and other depressions. These floodplains are often referred to as inland deltas. Some of the most important floodplains, such as the Inner Niger Delta in Mali, are in arid areas where their exceptional productivity is not only vital to the local economy, but also supports spectacular concentrations of waterbirds and other wildlife.

Top **The aptly-named Delta River in Alaska, USA.** The complex pattern of braided channels separated by mudbanks is typical of the lower reaches of rivers with a highly variable rate of flow. In this case, each spring thaw releases torrents of meltwater that flood the network of narrow channels. Sub-surface currents cut new channels in the river bed, and these are revealed when the flow decreases again in winter.

Above **Mangrove forest** cloaks the shores of the Khouran Strait in southern Iran. Mangrove leaves falling into the tidal waters are the basis of a food web that supports a rich ecosystem. The roots of the mangroves form a sheltered environment that provides a "nursery" for the young of many marine species. The fragility of the mangroves is underlined by the desert landscape of sand and rock visible in the distance.

Freshwater marshes

Stretching from Lake Okeechobee to the southwest tip of Florida, the Everglades covers over 7,000 square kilometres (2,700 square miles). It is one of the world's largest freshwater marshes. At the other extreme are the marshes that form in small, wet areas wherever groundwater, surface springs, streams or runoff cause frequent flooding, or create permanent areas of shallow water.

Although few freshwater marshes can compare in size to the Everglades, the vast number of small marshes makes this type of wetland among the most widespread and important worldwide. Large areas of southern Africa, for example, are dotted with "dambos", small freshwater marshes, which provide essential grazing and agricultural land for many rural communities. In North America, the Prairie Pothole region includes several million freshwater marshes at densities as high as 60 per square kilometre (150 per square mile) in some areas. Some of the larger marshes dominated by papyrus (*Cyperus papyrus*), cattail (*Typha* sp.) and reed (*Phragmites* sp.), and which have standing water throughout most of the year, are normally referred to as swamps.

Lakes

The diversity of lakes and ponds is the result of a host of different processes. Some lakes, such as Reelfoot Lake, Tennessee, in the United States, Lake Baikal, in the former Soviet Union and Lake Tanganyika, on the borders of Zaire, Tanzania, Burundi and Zambia, are formed by folding or faulting of the Earth's crust. Similarly, many crater lakes, including many in the Pacific islands, have been formed through volcanic disturbances. In the Northern Hemisphere, glacial action has been an especially important force. Cirque lakes, thaw lakes, and pothole or kettle lakes all owe their origin to the processes of glacial ice. The action and flow of rivers can also create a variety of lake types, such as oxbow and alluvial fan lakes, plunge pools and basins. Alpine lakes are formed by landslides and mudflows, while some lakes are remnants of larger ones, formed under more moist prehistoric environments. Shifting sediments by nearshore currents can create shoreline lakes cut off from larger seas of freshwater bodies.

Peatlands

Once thought to be restricted to the high latitudes of the Northern Hemisphere, peatlands are now known to exist on all continents and at all latitudes. They are even found in the tropics, where thick deposits form in association with marsh and swamp, particularly around lake margins and coastal regions. In total, peatlands are estimated to cover some 4 million square kilometres (1.5 million square miles). There is a great diversity of peatland worldwide, the pattern being governed by acidity, climate and hydrology (especially whether the peat is kept wet by direct rainfall or lateral groundwater flow). The highly distinctive, northern wetland landscapes of bog, moor, muskeg and fen are all examples of peatland.

In general terms, peat forms under conditions of low temperature, high acidity, low nutrient supply, waterlogging and oxygen deficiency. These specific circumstances slow the decomposition of dead plant matter. The characteristics of peatland ecosystems, however, are so varied that it is difficult to generalize on their functions and values. For example, some peatlands, namely bogs, are highly acidic and nutrient deficient, others, such as fens, are more or less neutral and rich in nutrients. Peatlands, therefore, include some of the least, as well as some of the most, productive wetlands.

Above **Spring in northern Finland. The zone between taiga and tundra habitats contains large areas of open marsh and bog, interspersed with stands of dwarf forest. During the winter months** groundwater is frozen solid. When spring sunlight melts the ice, small temporary ponds form. The plants that grow in the ponds provide the region's wildlife with a vital source of sodium.

Forested wetlands

Swamp forests develop in areas of still water around lake margins and in parts of floodplains, such as oxbow lakes, where water rests for long periods. Their character varies according to geographical location and environment. In the northern United States, for example, red maple (*Acer rubrum*), northern white cedar (*Fraxinus* sp.) and black spruce (*Picea mariana*) are prominent, while in the south, bald cypress (*Taxodium distichum*), black gum (*Nyssa sylvatica*), Atlantic white cedar (*Chamaecyparis thyoides*) and willows (*Salix* sp.) dominate.

In much of Southeast Asia the swamp forests are dominated by paper-bark trees (*Melaleuca* sp.) and other commercially valuable species. In Indonesia, swamp forests of this type cover over 170,000 square kilometres (65,000 square miles). Similar resources are found in the Amazon Basin. Here, the floodplains of the River Amazon and its tributaries support some of the most extensive flooded forests in the world. In Africa, the most extensive areas are found in the Congo Basin, where hundreds of thousands of square kilometres of the floodplain are densely forested.

Below **A swamp region in Australia. Rising water has drowned the trees in the centre of the picture. Their bases are now being colonized by grasses and other water-tolerant species.**

Right **Salt marsh along the estuary of the River Tow in Great Britain. Although not as productive as mangroves, this salt-resistant vegetation forms a vital part of estuaries in northern latitudes.**

How Wetlands Work

The hydrology of wetlands creates the unique conditions that distinguishes these environments from either terrestrial or deepwater habitats. Hydrological systems of different wetlands vary greatly in terms of frequency of flooding, and duration and depth of waterlogging. A coastal salt marsh typically floods twice a day, while also revolving around a monthly pattern of spring and neap tides. The water level between high and low tides also varies enormously, from being negligible in the case of the Gulf of Mexico or the Mediterranean, to more than 6 metres (20 feet) in many exposed estuaries, and more than 12 metres (40 feet) in special coastal configurations such as those of the Bay of Fundy in Canada. In strong contrast, the water level of many bogs and fens may remain consistently just below the peat surface throughout the year, dropping by a few centimetres or inches only during the summer or a drought period.

A very different pattern occurs on the world's major floodplains. Here, highly seasonal flood peaks give rise to floodwaters which commonly exceed 8 metres (26 feet). The precise pattern and timing of the flood depends on the season and location of rains and the shape of the drainage basin and floodplain.

It is the regularity of the flood pattern, however, which is so important in maintaining the structure and function of wetlands. Without a regular cycle, the productivity of fisheries, vegetation growth cycles and the success of wildlife migrations are seriously affected. In turn, the well-being of many human communities in Africa, Southeast Asia and South and Central America are likewise dependent on the flood pattern for the essential goods and services which the wetlands provide.

evaporation and transpiration

Left **A water buffalo, with its seemingly permanent companion the oxpecker, grazes in a seasonal freshwater marsh region in Northern Australia. The margins of the wetland support a variety of grasses. These types of wetland are important areas of grazing in many parts of the world.**

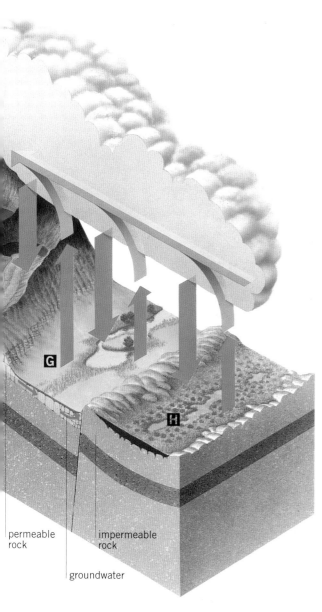

permeable
rock

impermeable
rock

groundwater

Right **Eutrophic stream
running through lowlying
deciduous woodland.
Under these near-stagnant
conditions, inflow of
nutrients stimulates algal
overgrowth, which in turn
deoxygenates the water.
Although this situation
was brought about by a
different set of conditions
to those in acidic bogs, the
result is the same – the
water supports few animals
or higher plants.**

Life-bearing water

Water movement is important in the transport of materials such as sediment, organic matter and nutrients to, from and within wetlands. Water, for example, provides a way for aquatic organisms to travel over what may well be dry land for significant periods of the year. This is vital for migratory fish, such as carp and bass, which have to cross floodplains to spawn in vegetation or swamp forest. And of course rivers themselves provide certain species of fish with the migration pathways they need.

The characteristics of wetlands are influenced strongly by water quality. Waters are classified according to their fertility, which in turn reflects nutrient content – from the lowest to the highest: ultra-oligotrophic, oligotrophic, mesotrophic, eutrophic and hypereutrophic. These terms apply primarily to lakes, where the extremes are easy to identify. At the oligotrophic end of the scale, waters are very clear, saturated with oxygen, but contain few nutrients essential to plant life, and therefore little animal life. Fast-running mountain streams are good examples of this condition. At the eutrophic extreme of the scale, waters are turbid, support dense algae growths, but have a very low oxygen content and therefore support few animals. Lowland enclosed lakes in agricultural catchments with low turnover of water and high nutrient inputs are characteristic of this condition.

The terminology can be extended generally to the wide range of wetland types. In the case of peatlands, for example, certain types of bogs are oligotrophic. The few nutrients they receive come solely through rainfall. Most fens, however, may be mesotrophic or even eutrophic. Fens usually obtain their nutrients from the lateral flow of groundwater, which carries the dissolved nutrients from soil or rock strata. Thus, while bogs are dependent on a certain amount of rainfall, there is no climatic restriction on the development of fens. The occurrence of fens is based on rock formation and the flow of groundwater. For this reason, fens exist in regions all over the world where it is too dry for the development of bogs.

Although bogs and fens both comprise accumulations of peat and may have similar water tables, the difference in water chemistry results in different vegetation and habitat for wildlife. Bogs are dominated by plants capable of withstanding acidic conditions, such as *Sphagnum* mosses, cotton grass (*Eriophorum* spp.), sedges like *Carex rostrata* and rushes like *Juncus squarrosus*. Insectivorous plants such as sundew (*Drosera* spp.) are also often present, indicating the need to supplement certain nutrients, particularly nitrogen, in such environments. The nutrient-enriched fens on the other hand support a rich variety of plants, including bulrush, common spike rush, bulbous rush, marsh pennywort and a wide variety of brown mosses.

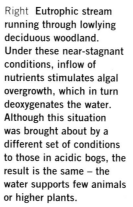

Above **The various types
of wetland shown in the
diagram have different
hydrological signatures.
Estuaries [A] and
mangroves [B], for
example, depend upon the
tides. As the tide changes,
the salinity of the water
can vary between almost
totally saline to freshwater.
Animals and plants that
inhabit these regions have
adapted to survive both the
daily flooding and drying
out as well as the salinity
variation. Floodplains [C]
and flooded forests [D]
tend to undergo seasonal
flooding. Their capacity
to store water can be
beneficial as they can
retain floodwater,
preventing flooding
downstream, and recharge
groundwater. This process
of recharging also purifies
the water as it slowly**

filters through sediment
and rock strata. Lakes
experience an ageing
process known as
eutrophication. "Old"
lakes [E] are characterized
by algal growth, a sign of
nutrient rich/oxygen poor
water. Fertilizer runoff
from fields can speed the
process. "Young" lakes [F]
tend to have clear, oxygen
rich/nutrient poor water.
Fens [G] and bogs [H]
differ in that fens
receive nutrients from
groundwater flow and can
sustain a wide diversity
of plants and animal life.
Bogs, on the other hand,
receive no groundwater
and are therefore very
acidic and lacking in
nutrients. These harsh
conditions are suitable
only for plants such as
Sphagnum moss, which
can tolerate acidic soils.

Succession

Wetlands evolve as a result of natural processes of development. Such change is led by successive waves of varying plant communities, which, as they develop, alter the environment. They produce habitat conditions increasingly less suitable for their own survival, but more advantageous for other species. This type of succession is known as *autogenic* succession. Another type of succession, in which change is caused by some external factor, such as a warming of the climate, is called *allogenic* succession.

The classic autogenic succession is illustrated in the development of a small freshwater lake or pond. The first flowering plants to colonize the wetland are floating species such as pondweeds (*Potamogeton* sp.) and duckweeds (*Lemna* sp.). Over time, these encourage detritus and sediment to build up around the wetland edges. The shallower water, in turn, allows the emergent species such as reeds and rushes to take root. These trap even more sediment and give way to terrestrial, shrub-like plants. In this way, the wetland is transformed to a dry-ground ecosystem by natural development. Succession can be accelerated and extended by human activities such as drainage, river diversion and groundwater abstraction.

Coastal succession

Considerable debate surrounds the existence of succession in coastal wetland systems such as mangroves. Early views were that species such as *Rhizophora* acted as pioneers, trapping sediment and acting as a land builder. This then allowed other mangroves such as *Avicennia* to establish themselves and eventually dominate the space previously occupied by the *Rhizophora*. In time, the pioneer species is forced seaward to survive, and in so doing progressively extends the landward margin. Such theories led to the term "land-builder" to describe the process of accelerated sedimentation and mangrove expansion. Rates of advance of more than 100 metres (330 feet) have been recorded and may lead to spectacular coastline changes. Palenberg, Indonesia, was once a thriving port on the north coast of Sumatra, visited by Marco Polo in 1292. Today, it is 50 kilometres (30 miles) inland.

More recent studies, however, indicate that while mangroves commonly exhibit zonation, there is by no means a uniform pattern and species distribution is determined by much more complex patterns of water level, salinity, sedimentation and flooding regime than previously recognized. So-called pioneers may simply respond to patterns and rates of sedimentation in the tropical coastal zone rather than control it. It may be much more accurate to consider mangroves as land consolidators or protectors rather than creators.

Wetlands and time

In terms of geological timescales, wetlands retain specific physical features for very short periods. This may be because they are associated with river valleys, floodplains or deltas, systems which are forever evolving new shapes or structures. Natural development or changes in river flow or direction, subsidence or erosion all cause rapid wetland degradation.

Coastlines are rarely static for significant periods of time and throughout the last 2 million years there have been major fluctuations of sea level. The latest rise, while certainly exacerbated by human activities contributing to the greenhouse effect, is actually a part of a long-established post-glacial (or interglacial) trend. Unless coastal wetland ecosystems can adjust to the rate of change of sea levels, such as through increased sedimentation, they will be subjected to a major alteration of their water cycles, which may result in longer periods of inundation, saltwater stress, exposure to different, and potentially damaging, animal species and increased erosion. If the coastal

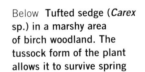

present lake bed

original lake bed

Below **Tufted sedge (*Carex* sp.) in a marshy area of birch woodland. The tussock form of the plant allows it to survive spring flooding. By raising the green parts up on the tussock, the plant can keep growing even at the height of the floods (marked by the transition from dead to live leaves).**

Left Succession of a pond or lake illustrates the pioneering aspect of particular species of wetland plants. Floating species, such as water lilies, encourage detritus and sediment build-up, which over time (represented by the arrows) enables emergents, such as reeds and rushes, to take root. Further sediment build-up occurs, creating a sufficiently firm substrate on which terrestrial plants can eventually grow.

Right White water lilies (*Nymphaea alba*) growing in southern Britain. Although it is classed as a floating plant, the water lily is in fact firmly rooted. Encased in the bottom mud, a thick rootstock sends stalks up to the surface which terminate either in circular leaves or a single flower. The leaf stalks and flower stalks both carry longitudinal air passages. The family Nymphaeaceae have a worldwide distribution, and include the giant water lily (*Victoria regia*) with leaves that measure up to 2 m (6 ft) in diameter. This particular species is native to Guyana and Brazil.

Palenberg in Indonesia was once a thriving port on the north coast of Sumatra, and was visited by Marco Polo in 1292. Today, the former port lies 50 kilometres (30 miles) inland.

topography is suitable the wetland fringe may simply retreat, but if gradients are too steep, or if artificial flood defences are put in place, the result will be progressive loss of salt marshes, mud flat complexes and mangroves.

Changing climate

Wetlands are highly sensitive to climatic change. This is particularly true of the world's peatlands, which cover large tracts of the Northern Hemisphere. The blanket bog and raised bogs of northwest Europe are particularly vulnerable to amount and frequency of rainfall. Any reduction in the precipitation/ evaporation ratio is likely to lead to increased decay and eventual wastage of the peat mass. In the case of the vast peatlands of the tundra and taiga, a drop in the water table can result in increased carbon release, so contributing further to the greenhouse effect and global warming potential.

Irrigated fields beside the Indus River in northern India. The flat plain and braided channels of the river indicate seasonal flooding, which in this case is fed by spring meltwater from the southern slopes of the Himalayan mountain range. The natural floodplain landscape, with its diversity of wetland habitats, and plant and animal species, has been transformed by thousands of years of human occupation and agriculture.

Why We Need Wetlands

For 6,000 years, river valleys and their associated floodplains have served as centres of human population, with many boasting sophisticated urban cultures. Their fertile soils brought in huge harvests upon which the peoples of the regions could depend. Today, the wetlands which nurtured the great civilizations of Mesopotamia and Egypt, and of the Niger, Indus and Mekong valleys, continue to be essential to the health, welfare and safety of millions of people who live by them.

All wetlands are made up of a mixture of soils, water, plants and animals. The biological interactions between these elements allow wetlands to perform certain functions and generate healthy wildlife, fisheries and forest resources. The combination of these functions and products, together with the value placed upon biological diversity and the cultural values of certain wetlands, makes these ecosystems invaluable to people all over the world. Within this section, wetland values are introduced in general terms. The Atlas section looks at specific wetland uses, illustrating the immensely varied wetland resources and values upon which people depend.

Flood control, water purification and shoreline stabilization, for example, can all be maintained by wetland systems. However, most wetland development projects today concentrate intensively on one aspect, such as the agricultural or fishery yield. The limitations of this approach are now becoming obvious. If wetlands are converted or developed without first taking into consideration their full value, the negative consequences can be felt immediately by local people. The economy of a region, or indeed a nation, may be affected adversely if the alterations are many or large. Development often requires major investments of capital, manpower, technology, and inputs such as fertilizer, as well as substantial annual investments in maintenance. And where conversion is attempted, the ability of natural wetlands to sustain alternative development is found to be low; moreover, such conversions require far more sophisticated management than is generally available to the average rural community.

Despite the importance of the range of resources and services which wetlands provide, we have tended to take these for granted. As a result, the maintenance of natural wetlands has received low priority in most countries. But even as apathy and ignorance continue to permit conversion of wetlands, people are becoming increasingly aware of the loss of the services wetlands once provided free of charge.

Groundwater and flood control

When water moves from a wetland down into an underground aquifer (a rock deposit that contains water), it is said to recharge groundwater. By the time it reaches the aquifer, the water is usually cleaner, due to filtering processes, than it was on the surface. Recharge is also beneficial for flood storage because runoff is temporarily stored underground rather than moving swiftly downstream and overflowing.

Below **Aftermath of flooding in Khartoum, Sudan. Seasonal variations in the level of the River Nile are well known, and local people avoid building in areas liable to flooding. Recent famines have caused the population of Khartoum to rise sharply under an influx of refugees from the country. Pressure of space has made these newcomers occupy the lowest ground with predictable results.**

Right **If floodplains are preserved [A], when floods occur [B], the floodwaters are free to spread out over the plain, causing little damage downstream. If, however, floodplains are developed [C], the floodwaters are channelled downstream, where the risk of damage is far more severe – caused not only by the flood heights, but also by the velocity of the floodwater as it flows downstream [D].**

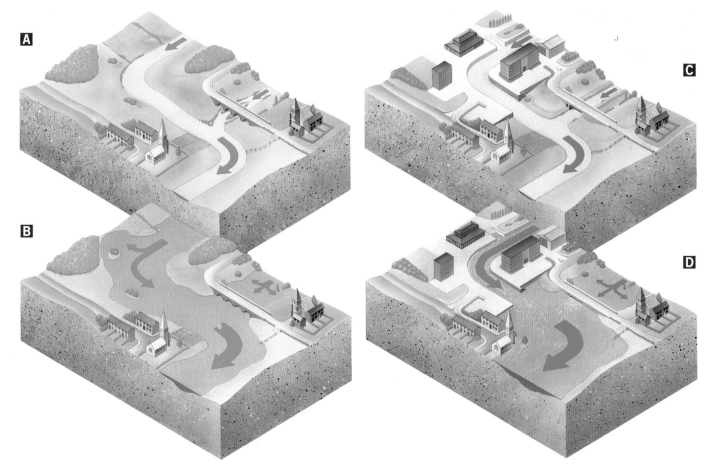

Once in the aquifer, water may be drawn out for human consumption through wells, or it may flow laterally underground until it rises to the surface in another wetland as groundwater discharge. In this way, recharge in one wetland is linked to discharge in another. Wetlands that receive most of their water from groundwater discharge usually support more stable biological communities, because water temperatures and levels do not fluctuate as much as in wetlands that are dependent upon surface flow.

By storing rain and melted snow and releasing runoff evenly, wetlands can diminish the destructive onslaught of floods downstream. Preserving natural storage can avoid the costly construction of dams and reservoirs. In the Charles River of Massachusetts in the United States preservation of 38 square kilometres (15 square miles) of mainstream wetlands provides natural valley storage of flood waters. It is estimated that had 40 per cent of these wetlands been reclaimed, the increased flood damage would have cost US$3 million each year. And had they been filled completely, the added flood damage would have been over US$17 million per year.

Stable shores and storm protection

Wetland vegetation can stabilize shorelines by reducing the energy of waves, currents or other erosive forces. At the same time, the roots of wetland plants hold the bottom sediment in place, preventing the erosion of valuable agricultural and residential land, and property damage. In some cases, wetlands may actually help to build up land. To illustrate the effect of wetlands in protecting against destabilizing sea forces, it was estimated in the United Kingdom in 1981, that sea walls constructed behind salt marshes would be over 20 times cheaper to build than walls unprotected by salt marshes.

Hurricanes and other coastal storms cause wind damage and flooding. In the developed world, the principal consequence is property damage; in poor tropical nations, however, it is more often human injury and death. In Bangladesh, 150,000–300,000 people were killed in one storm surge in 1970. Many wetlands, in particular mangroves and other forested coastal wetlands, help dissipate the force and lessen the impact of coastal storms. The mangrove

forest of the Sundarbans in India and Bangladesh, for example, breaks storm waves which often exceed 4 metres (12 feet) in height. In recognition of their protective functions, over the last 10 years the Bangladesh Government has planted vast areas of mangroves to protect embankments and farms.

Sediments and nutrients

Sediment is often the major water pollutant in many river systems. Many wetlands help reduce this by serving as pools where sediment can settle. In addition, if reeds and grasses are present to slow down a river's flow, the opportunity for settling is increased.

Although the build-up of too much sediment in a wetland may alter its biological functions, such as floodwater storage and groundwater exchange, the quality of ecosystems downstream can be improved if suspended sediment is held in the headwaters (tributaries located near the source of a river). Because toxic substances, such as pesticides, often adhere to suspended sediment, they too may be retained with the sediment. Retaining sediment in headwater wetlands can lengthen the lifespan of downstream reservoirs and channels, and reduce the need for costly removal of accumulated sediment from dams, locks, power-stations and other man-made structures.

Nutrient retention and export

Wetlands retain nutrients, most importantly nitrogen and phosphorus, by accumulation in the sub-soil, or storage in the vegetation itself. Wetlands that remove nutrients improve water quality and help prevent eutrophication – a process, whereby the build-up of excess nutrients leads to rapid plant growth, increased oxygen demand and ultimately reduced productivity and biological diversity. By removing nutrients wetlands can dispense with the need to build water treatment facilities. Under certain circumstances, wetlands can even be used for treatment of domestic waste from small, non-industrial communities.

When wetlands remove nutrients (and pollutants) they are referred to as "sinks". This is particularly important with regard to nitrates which can be converted back to harmless nitrogen gas and returned to the atmosphere. When nutrients are returned to the surroundings, wetlands are said to act as "sources". A common role of wetlands during the growing season is to accumulate nutrients when the water flows slowly. These nutrients support fish and shrimps, as well as the forest, wildlife and agricultural wetland products. When water flows fast, wetlands may act as a source.

This cycle has important implications for algal growth, water quality, fish production and recreation downstream from the wetland area because it reduces nutrient levels at a season of the year when added nutrients are likely to cause eutrophic conditions downstream. Release of the nutrients occurs when they are less likely to cause eutrophication.

Many wetlands support dense populations of fish, cattle or wildlife which feed on the nutrient-rich waters or graze on the lush pastures. But in addition

Above **The villagers of Dixcove in Ghana, West Africa, make full use of their wetlands. The village is situated on the banks of a tidal estuary, and there are rich fishing grounds literally on the doorstep. Far from being a barrier to human activity, the river estuary provides the basis for human existence.**

Top left **Young mangrove trees planted as part of a shoreline reclamation project in Miami, Florida, USA. The tree roots will stabilize the mud and encourage sediment to accumulate. An arc of boulders has been placed to protect the saplings from the wake produced by motor-boats.**

Right **Graceful progress across a Burmese lake. Wetlands as transport systems can be overlooked by city dwellers. In the rural districts of many developing countries, water transport often is the cheapest and most reliable method of carrying people and goods. In areas of permanent wetland, boats may be the only practical means of transport.**

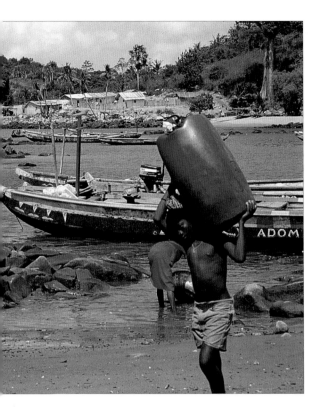

to this production within the wetlands, environments downstream and in coastal waters also benefit from nutrients that are carried away by surface flow, in streams, or by groundwater recharge.

In temperate regions, nutrients stored in growing wetland plants are released when the water cools and the plants die in winter. Part of the value of river and coastal fisheries is attributable to this vital support function provided by wetlands, and is an addition to their role as breeding or nursery areas for fish. Interfering with wetlands can disrupt nutrient production. After construction of the Aswan Dam, for example, on the River Nile, export of nutrients and sediments was substantially reduced. As a result, total fish catch from the Mediterranean Sea adjacent to the Nile Delta decreased from 38,000 tonnes in 1962 to 14,000 tonnes in 1968.

Further uses and resources

If our harvesting of wetland plants and animals respects the annual production rates and regenerative capacity of each species, we can enjoy the benefits of wetland productivity without destroying an important habitat. For example, direct harvest of the forest resources of many wetlands yields a number of products, ranging from fuelwood, timber and bark, among wood products, to resins and medicines which are common non-wood, forest products.

Wetlands that contain important grasslands grazed by livestock are important to local communities. The Brazilian Pantanal, for example, supports over 5 million cattle. And over much of Africa, wetland grasslands provide a critical source of dry-season grazing. Leaves, grasses, and seed pods may also

be collected as fodder for sale, or used as a dry-season cattle feed.

Two-thirds of the fish we eat depend upon wetlands at some stage in their life cycle. In Africa, fish are the most important source of animal protein, making up 20 per cent in 1980–1982. Many of the fish are caught in small ponds and marshes on a daily basis. In other wetlands, such as the Inner Niger Delta in Mali, there are annual fishing festivals towards the end of the dry season.

Wetland wildlife also provides an important source of protein and income even in the industrialized world. In Canada, the value of mink, beaver and muskrat exceeded US$43 million in 1976. Further south, in El Salvador, 30,000 tree duck eggs were harvested at Laguna Tocotal between 1981 and 1985, providing an important protein resource for the local communities.

While many wetlands have been converted to intensive agriculture, many others, however, continue to be cultivated in their natural form. Properly managed, natural wetland agriculture can often yield substantial benefits to rural communities. Indonesian tidal swamps have been cultivated successfully by communities using traditional techniques. Rice is the major crop, often in combination with coconuts and fruit trees, which help reduce soil acidity. Corn, cassava and vegetables are also grown. In addition to these crops, swamp farmers also raise livestock (principally poultry), and maintain fisheries in the canals and ditches of the coconut gardens.

Wetlands are also important as a genetic reservoir for certain species of plant. Rice, a common wetland plant, is the staple diet of over half of the world's people.

Left **Cooter turtles** (*Pseudemys floridana*) and green-backed heron (*Butorides striatus*) in Florida, USA. The wide variety of wetlands has led to the evolution of many specialized species. Second only to rain forests, wetlands contain a greater diversity of wildlife than any other terrestrial habitat, and a higher proportion of endangered species.

Below **Cutting peat in Caithness, Scotland. These vast peatlands represent a virtually inexhaustible supply of fuel for scattered communities. Small-scale use does not harm the environment, but in other places peat is mined extensively and put to industrial uses, adding to greenhouse gases. Ireland generates 40 per cent of its electricity from seven peat-fired power stations.**

Energy resource

Some wetlands contain potential energy for human consumption, normally in the form of plant matter and peat. When this can be used on a sustainable basis, it is an important component of an integrated management scheme for the wetland. However, when extraction of peat is carried out on a large scale, there is a real danger of it destroying the ecosystem. Proposals for peat mining in Jamaica, Brazil and Indonesia, for example, have aroused considerable international concern in recent years.

Biological Diversity

Many wetlands support spectacular concentrations of wildlife. In West Africa, the floodplains in the Senegal, Niger and Chad basins support over a million waterfowl, many of them migratory, visiting the region during the European winter. And in Mauritania the tidal flats of the Banc d'Arguin National Park provide a wintering site for some 3 million shorebirds each winter. In Zambia, 30,000 black lechwe antelope (*Kobus leche smithemani*) inhabit the Bangweulu Basin, along with one of Africa's most important populations of sitatunga (*Tragelaphus spekei*) and shoebill storks (*Balaeniceps rex*). In Brazil, the Pantanal covers over 100,000 square kilometres (39,000 square miles), with large populations of spectacled caiman (*Caiman crocodilus*), capybara (*Hydrochoerus hydrochaeris*) and jaguar (*Panthera onca*), as well as one of the most distinctive mosaics of vegetation in Latin America.

While it is concentrations of individual species, rather than the diversity of species that has attracted most attention from conservation scientists, many wetlands do support a significant diversity of vertebrates, many of which are endemic (unique to the area) or endangered. The diversity of fish species is notable. For example, in East Africa, Lakes Victoria, Tanganyika and Malawi support over 700 species of endemic fish. In Lake Tanganyika alone, 214 species have been identified, 80 per cent of which are endemic.

In many countries the inaccessibility of wetlands has attracted species which, although not confined to wetlands, are dependent upon the shelter they provide. In India and Bangladesh, the mangrove forest of the Sundarbans is the largest remaining habitat of the Bengal tiger (*Panthera tigris*). Over much of Latin America, wetlands such as the Pantanal of Brazil, Paraguay and Bolivia, and the Mesquitio lowlands of Nicaragua and Honduras provide the most important habitat for the jaguar.

Wetlands are also important as a genetic "reservoir" for certain species of plant. Rice, a common wetland plant, is the staple diet of over half of the world's people. Wild rice in wetlands continues to be an important source of new genetic material used in developing disease resistance and other desirable traits.

Left **A party of tourists in Big Cypress Swamp of the Florida Everglades. About a million tourists visit the Everglades each year, making it one of the United States' top** attractions. Like all environments, wetlands must compromise with tourism. Supervised tours such as this do little harm; but uncontrolled tourism can cause serious damage.

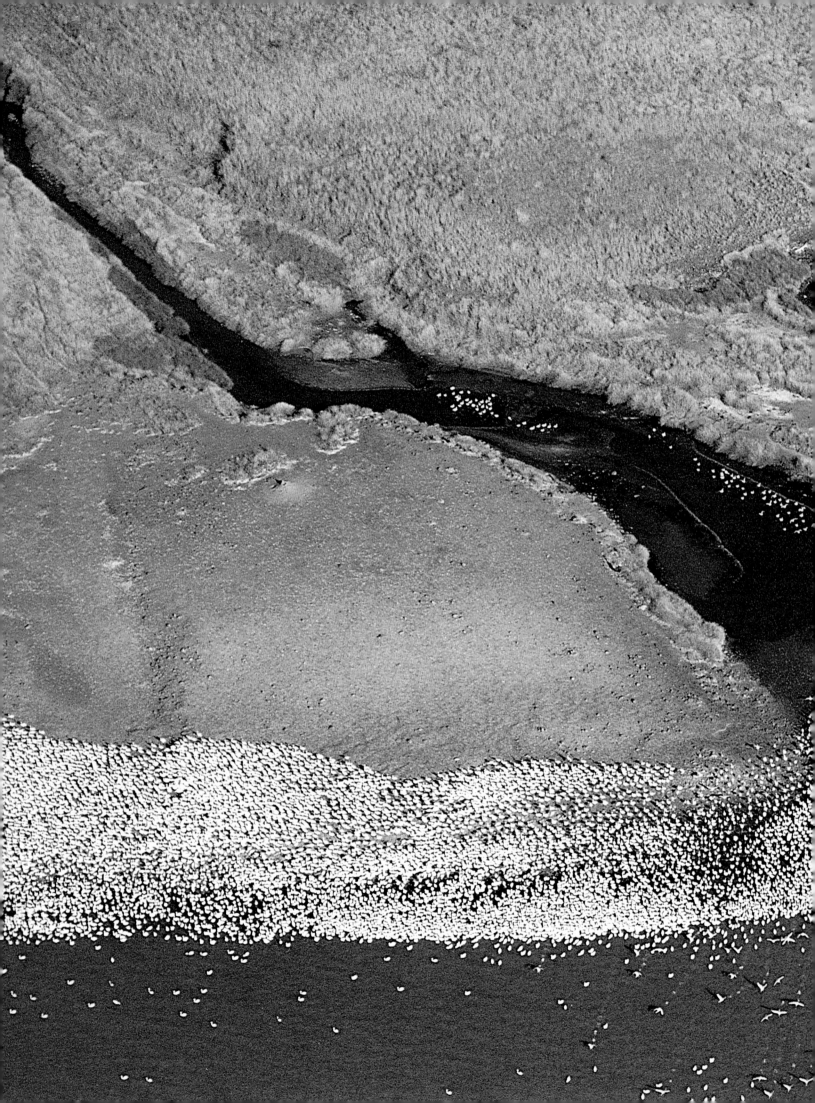

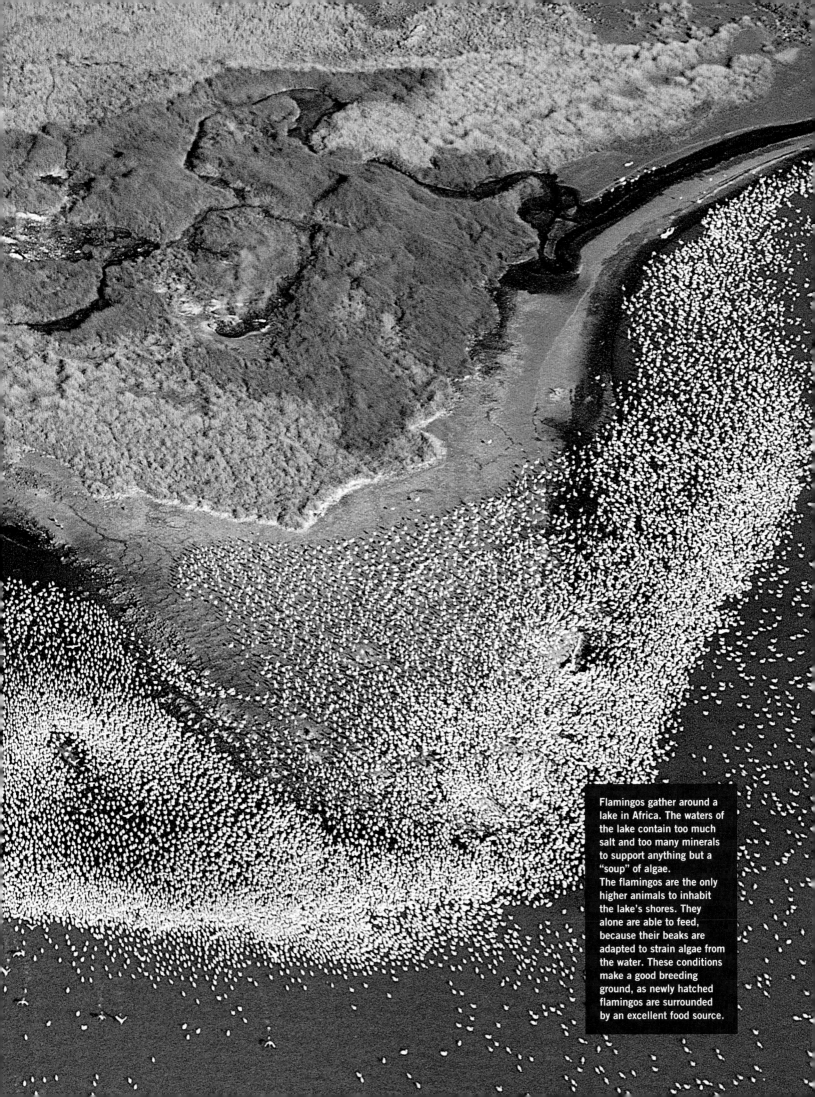

Flamingos gather around a lake in Africa. The waters of the lake contain too much salt and too many minerals to support anything but a "soup" of algae. The flamingos are the only higher animals to inhabit the lake's shores. They alone are able to feed, because their beaks are adapted to strain algae from the water. These conditions make a good breeding ground, as newly hatched flamingos are surrounded by an excellent food source.

Adapting to Life in Wetlands

Many wetlands are lush habitats that provide countless species of plants and animals with a rich environment from which they can obtain most of their requirements. However, as strange as it may sound, the relationship between wetlands and water is not always simple and uniform. Plants and animals living there must adapt in order to survive alternate periods of flooding and desiccation, and the consequences this lack of uniformity brings.

Plant Adaptation

Plants have developed a wide range of adaptations in order to survive and exploit their wetland environments. The most obvious developments are structural, concerned especially with the problem of supplying oxygen to roots growing in anaerobic (oxygen-deficient) soils or sediments. Most aquatic plants, such as water lilies (Pontederiacceae family), are extremely porous and contain special tissue, called *aerenchyma*, which has large, air-filled intercellular spaces. Oxygen diffuses 10,000 times faster in air compared to water, and so the aerenchyma are thought to facilitate the movement of oxygen from the leaves to the root-like rhizomes.

Mangrove "knees"

Some trees, notably mangroves, such as *Avicennia* species and the swamp cypress (*Taxodium distichum*), have evolved curious projections called *pneumatophores*, or in the case of the swamp cypress, woody "knees". They develop from lateral roots growing near to the surface, and protrude vertically up to 30 centimetres (12 inches) above the soil or sediment surface. The precise function of these organs has been the subject of considerable scientific debate, but there is general agreement that they assist the plant in maintaining adequate root respiration. The root structures often produce dense networks which may also help to accumulate and stabilize sediment, providing a firm base on which the tree can grow.

Rather less prominent, but no less important, are the enlarged pores, called *lenticels*, which typify mangroves such as *Pelliciera* species. They are found on the bark and allow gaseous exchange between the atmosphere and the mangrove's tissue in the more or less continuously waterlogged environment of coastal swamps.

Plant chemistry

The external and internal adaptations which facilitate the movement of gases between plants and the atmosphere are important to only some wetland species. Others have developed biochemical adaptations, which scientists now think may be of much greater significance in terms of tolerating flood or waterlogged conditions. These complex biochemical adaptations work in one of two ways. Some species oxidize toxic elements, such as ions of iron (in the form Fe^{2+}), which diffuse into the root from the soil, altering their chemical structure and rendering them harmless. This oxidation process is thought to be highly efficient on the large internal and well-aerated surface of the aerenchyma tissue.

Below **Common sedge (*Carex nigra*) in flower. The 4,000 species of sedge all favour damp marshy ground, and will flourish even if their roots are entirely submerged.**

Right **Bald Cypress trees in Okefenokee Wildlife Refuge, Georgia, USA. Bald Cypress has adapted to its habitat by evolving huge "knee" structures which stand above the water. Although these probably help provide oxygen to the roots, they are also thought to add stability as the trees grow to 50 m (160 ft).**

Bottom **A sedge and mare's tails (*Hippuris vulgaris*) are among the emergent plants that rise above the waters of a marsh. Air passages in the stems supply oxygen to the submerged roots. This adaptation was probably the first terrestrial plants evolved under similarly damp conditions about 400 million years ago.**

Other species of plants have developed mechanisms that excrete the toxic products of anaerobic respiration. Some, for example, diffuse acetaldehyde and alcohol, the toxic by-products, through the large surface area provided by finely divided roots. Other species, such as willow (*Salix* spp.) are capable of immobilizing or converting the harmful by-products to less toxic forms, such as, in the willow's case, pyruvic and glycolic acid.

Emergent plants
Rushes, reeds, sedges and some grasses typify wetland plants that are firmly rooted in the soil or sediment, but have erect stems that grow out of the water. Thus, while the stem is exposed to the atmosphere in the same way as any fully terrestrial plant, the roots have to cope with anaerobic conditions and, in some cases, salty environments. As with most wetland species air spaces in roots and stems allow oxygen to travel from the aerial parts of the plant to the roots. While land plants may have a porosity of about 2–7 per cent of volume, up to 60 per cent of many wetland plants is pore space.

Salt marsh species and mangroves also face the problem of salts. Some mangroves, including *Rhizophora* species, have specialized cells in the roots which block sodium while allowing essential elements, such as potassium, to move freely about the plant. Other species do not block salt at the roots, but have mechanisms for secretion. This is well illustrated in salt marsh grasses, such as *Spartina* species and some mangroves, the leaves of which are often covered with crystalline salt particles secreted from specialized glands.

Fuel efficiency

For the vast majority of plants, the first step of carbon fixation – part of the complex chemical process of photosynthesis in which carbon dioxide is converted into carbohydrates – is the production of a compound called phosphoglyceric acid. This is a three-carbon-atom compound, hence these plants are known as "C_3" plants. Many wetland plants, however, such as *Spartina* and *Panicum* species produce a four-carbon-atom compound called oxaloacetic acid.

Scientists have discovered that so-called C_4 plants have a much lower photorespiration rate than C_3 plants. And because photorespiration (light-stimulated respiration) is essentially a wasteful process (under some conditions, 30 per cent of carbon reduced during photosynthesis can be reoxidized to carbon dioxide during photorespiration) C_4 plants are much more efficient than C_3 plants both in rate of carbon fixation and in water use. One of the benefits of efficient water utilization is that it reduces the rate at which soil toxins, present in generally high concentration in the oxygen-deficient wetlands, are drawn into the root network. This increases the opportunity for detoxification to occur in the thin oxygenated zone around the roots.

Successful reproduction

Some plants, such as roses, reproduce sexually. Others, such as strawberries, reproduce vegetatively, sending out "runners" which take root and form a "daughter" plant. Papyrus (*Cyperus papyrus*) can spread by sexual as well as

> The structure of *Sphagnum* mosses enables them to hold enormous quantities of surface water, sometimes exceeding up to 15 times the weight of the plant itself.

Right **The carnivorous adaptation of the sundew (*Drosera rotundifolia*) is a response to the low levels of nutrients common to some wetlands. Insects are trapped by sticky droplets on the ends of "tentacles" found on the leaves. Enzymes liquefy the prey, allowing nitrogenous compounds to be absorbed into the plant body.**

vegetative means and is one of the fastest growing plants on Earth. The numerous spiky flower heads of papyrus produce thousands of seeds which fall into the river. Dispersal is controlled by the speed and extent of the downstream flood. Inevitably, some of the seeds become embedded in the sediment and germinate. While the papyrus' sexual reproductive method colonizes new areas, its capacity for vegetative reproduction ensures older communities remain stable. New shoots are sent up from the rhizomes at regular intervals. Within the first three months they have grown, matured and died – but as one shoot dies the nutrients are withdrawn and passed on to the next generation in a continuous process of development. The efficient conservation of nutrients and its extraordinary rate of growth contributes to the plant's great success in many African swamps.

Left **Spraying herbicide to control water hyacinth in the Sudan. This plant thrives in wetlands where pollution has increased nutrient levels.**

Below *Sphagnum* **moss showing berry-like fruit capsules.** *Sphagnum* **is one of the few plants that can tolerate the acidic conditions found in bogs.**

Peat-formers

The bog mosses (*Sphagnum* spp.) – also known as peat mosses – are exceptional wetland plants. They are capable of growing in very acidic conditions where, because there is no groundwater flow, the only supply of nutrients is from rainfall. Their structure enables them to hold enormous quantities of surface water, sometimes exceeding up to 15 times the weight of the plant itself. In this way a layer of *Sphagnum* moss can maintain high levels of ground saturation even during prolonged dry periods. As the plant continues to grow upwards, the lower parts die and become progressively compressed under the weight of accumulating vegetation. The slow rate of decomposition, thanks to the acidic conditions, makes *Sphagnum* an exceptional peat-former and maintainer. These species are particularly characteristic of the raised bogs, blanket mires and muskeg terrains of the oceanic, high-altitude and high-latitude regions.

Surface floaters

Many wetland plants are surface-floating species. Apart from the well-known water lilies (*Nymphaea* and *Nuphar* spp.), these include the water lettuce (*Pistia stratiotes*), water fern (*Azolla nilotica*), numerous members of the duckweed family (Lemnaceae) and water hyacinth (*Eichhornia crassipes*). Some, like the water hyacinth, have high growth rates. A native of South America, the introduction of water hyacinth to other subtropical and tropical regions, including the southern United States (where it is known as the "Florida devil"), India ("Bengal terror"), Africa and Southeast Asia, has enabled this adaptive plant to exploit new territory, causing problems of weed control.

Water hyacinth grows rapidly into extensive surface mats especially in sheltered inlets or oxbows. Floods, wind and wave action can break up the mat, causing it to spread rapidly in rivers and lakes. Floating mats can become so large that the force of the vegetation is sufficient to damage man-made structures. Navigation is sometimes completely impossible and drainage may be impeded. Where water hyacinth has hampered drainage in parts of Africa, habitat conditions are often created that are ideal for carriers of disease such as bilharzia-carrying snails and malaria-transmitting mosquitoes.

Although the waters infested by excessive growth of water hyacinth become depleted in oxygen, there can be some benefits. The plant is a highly effective absorbent of nutrients, such as nitrogen and phosphorus, together with contaminants and other toxic wastes. In Southeast Asia, the water hyacinth is used as a cattle food, providing an important source of income.

Much attention has focused on water hyacinth together with other floating aquatic plants such as *Lemna* and *Azolla*. The relatively high-nutrient content of the plants, especially in eutrophic waters, make them excellent fertilizers. They have been used to great effect in Sudan, India, Malaysia and China.

Animal Adaptation

Many animals, like plants, have adapted well to an aquatic existence. Their adaptations have naturally centred around breeding and feeding in a wetland environment. To exploit the many rich wetland habitats, such as estuaries and mud flats, animals have evolved their own unique adaptations to ensure that they can exploit one particular niche more effectively than their competitors.

Insect adaptations

To survive in wetlands, insects have developed a host of adaptations. Some, such as dragonflies and damselflies, for example, lay their eggs in the tissues of submerged plants. Water boatmen (family Corixidae) and back-swimmers (family Notonectidae) have developed strong legs fringed with hairs that propel them through the water as oars do a boat.
Many aquatic insects, including mosquitoes, when at their larval stage, have tiny tubes attached to the abdomen that poke out of the water, allowing the larvae to breathe. Others, such as the great diving beetle (*Dytiscus marginalis*), breathe by trapping air bubbles under their wings before each hunting dive. The larvae of some species, such as caddisflies (*Limnephilus* spp.), obtain oxygen from the water itself. They absorb the oxygen through thin sections of their fine outer body casing, known as the cuticle.

Water beetles and water boatmen are highly efficient hunters, pursuing their prey through the underwater jungle with speed, grace and agility. Other insects have adapted to life on the surface film, which provides a rich hunting ground for many land animals falling into the water. The pond skater (*Gerris lacustris*) has specialized legs that enable the insect to literally skate over the surface of the water. The limbs' lower segments are wide and paddle-shaped, with hairs that repel water and increase the surface area. Whirligig beetles (family Gyrinidae) also swim on the surface, but unlike the pond skater, they are half submerged. Their eyes have compensated for this half-in, half-out arrangement by being divided into two, allowing them to see both above and below the surface of the water. Both groups of insects can quickly cover large areas of stillwater, scavenging the unfortunate animals unable to escape the force of surface tension.

Crustaceans and carnivores

Among invertebrates, it is the crustaceans that show some of the most sophisticated adaptations to aquatic life. The water fleas *Daphnia* are highly mobile free-living crustaceans that inhabit shallow ponds and ditches. They have developed appendages specially adapted for filter-feeding. Close-set rows of long, fine, feathery bristles cover the limbs attached to the thorax (the middle section of an invertebrate). The bristles remove coarse particles from detritus or plankton and pass them backwards to be either swept away with the water current or broken up for food by the limbs further back along the body. The gathered material is passed towards the mouth via a "food groove" which runs along the flea's underside.

A rich variety of carnivores exists even among the invertebrates. As well as the many diving beetles, pond skaters and water bugs, there is one particular group, the dragonflies, whose nymphs have evolved a highly efficient weapon. They lie concealed in vegetation and seize passing prey, such as waterfleas or larvae of other organisms, with a "mask" that shoots out on a hinged arm and grabs the victim between two sharp hooks.

Below **Pond skaters feeding on a dragonfly. Adaptations to the ends of the skaters' legs enable them to stand on the water surface without breaking the "skin" formed by surface tension. Standing on the surface enables the skaters to detect ripples produced by heavier and less well-adapted prey struggling in the water. By homing in on the source of the ripples, the skaters are assured of a meal.**

Below **Although mosquito larvae (*Culex pipiens*) have an aquatic lifestyle, they have no specialized organs for breathing underwater. In order to take in oxygen, most species pierce the surface with a thin tube that grows from the abdomen. The "grip" of surface tension is sufficient to hold the animals at the surface, so that they can manoeuvre to take in food particles while breathing.**

Bottom **Common frogs spawning. Amphibians are ideally suited to wetlands. While some remain in the water all their lives, others return only to breed. Amphibians use external fertilization – eggs and sperm are deposited separately in the water, where fertilization takes place. After hatching, the young breathe through gills. On reaching maturity, most species lose their gills and develop lungs.**

Amphibians

Water is essential for amphibians. Although they can spend much of their time on land, amphibians, because their eggs are unprotected against drying out, must return to the water to breed. Tailed amphibians include the newts and salamanders. The largest European newt is the crested newt (*Triturus cristatus*), which is found in slow-moving waters with dense aquatic vegetation. The female lays between 200 and 400 eggs singly on the submerged leaves of grasses or aquatic plants. The eggs hatch after two to three weeks. The larvae (tadpoles) have external gills and live in water for some three months before metamorphosis is completed. The adults live on dry land outside the breeding season, but utilize the cover of stones, wood or moss and feed on worms, snails and insect larvae. Young newts may not mature for two to three years and do not return to the water to breed before then.

Frogs and toads make up the group of tailless amphibians. They are found in wetlands throughout the world, and occupy a wide range of altitudes, water salinities and qualities, as well as vegetation types. Some, such as the edible and marsh frogs (*Rana esculenta* and *R. ridibunda*), are found mainly in low-lying wetlands; the common frog (*R. temporaria*) lives in mountainous regions of Europe and Asia. The moor frog (*R. arvalis*) thrives in peat and swamp conditions, and ranges from northwest France to Siberia, as far as the Arctic Circle.

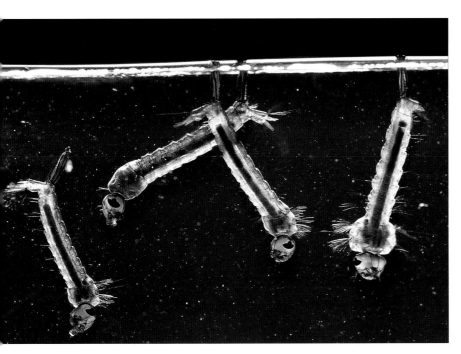

Tree frogs (family Hylidae) such as the common tree frog (*Hyla arborea*) live in lush vegetation in most tropical and subtropical regions of the world except Africa. These frogs have specialized hands and fingers that have circular pads from which is secreted a gluey substance. This helps them cling to the slippery leaves of trees and shrubs, where they capture insects often by means of a distinct leaping action. Tree frogs worldwide are typically small and vivid green in colour. Their coloration acts as a camouflage, matching the foliage with which they are normally associated.

Reptiles

Reptiles such as snakes, iguanas and crocodiles are common in the world's subtropical and tropical wetlands. Some, such as the grass snake (*Natrix natrix*), also live in the high-latitude wetlands of the Northern Hemisphere. Other species of *Natrix*, such as the Dice snake (*N. tessellata*), are well adapted to wetlands, being extremely proficient swimmers and particularly adept at catching fish. Active by day as well as night, the snake is found in freshwater and wetland edges from Europe to Asia. Like the grass snake, however, it needs to lay its eggs on land and so is dependent on a habitat where land lies close to its watery hunting ground.

Some wetland snakes, such as the cottonmouth moccasin (*Agkistrodon piscivorus*) with its distinctive white inner mouth, are highly poisonous. This particularly dangerous species of snake inhabits the cypress swamps of the southern United States and feeds primarily on frogs, small fish, salamanders and crayfish.

Crocodiles and alligators

These are the world's largest reptiles and are a key feature of wetlands, from the warm subtropical regions of the Americas to the tropics of Africa, Southeast Asia and Australia. Unlike most animals, crocodilians continue to grow throughout their lives – some males grow to over 4.5 metres (13 feet) and weigh over 225 kilograms (500 pounds). It is thought that crocodilians can live to be over 100 years old and for this reason they represent a force for considerable stability within the wetland wildlife community.

Crocodilians are effective hunters. They can move surprisingly quickly on land, and the whipping motion of their long, muscular tails makes them efficient swimmers. Bony plates

> It is thought that crocodilians can live to be over 100 years old, and for this reason they represent a force for considerable stability within the wetland wildlife community.

Below **Despite its name, the English grass snake has decidedly aquatic habits, and is frequently found in damp and marshy areas. The snake is an excellent swimmer, and frogs constitute a major part of its diet. Its European relative *Natrix viperinus* is even more aquatic and feeds almost exclusively on fishes. The North American moccasins (*Natrix fasciatus*) have a similar semi-aquatic lifestyle, as do the North American garter snakes (*Thamnophis* sp.)**

(osteoderms) provide an armoured shield which protects them from virtually all predators. With eyes and nostrils located on top of their heads, crocodilians can breathe and look around with only a small part of their body protruding from the water. This enables them to approach shoreline animals without being seen. Once the jaws take hold of the prey, release is virtually impossible and the animal is then usually dragged under water and drowned. The muscles which close the jaw are extremely strong and the array of teeth can be replaced when worn out or lost. A single animal may go through 3,000 teeth in its lifetime.

Hard times

The water cycle is as critically important to alligators and crocodiles as to all animals of the wetland habitat. Alligators in southern Florida excavate the marl (a fine-grained clay or loam rock) and soft, weathered limestone bedrock to create depressions commonly called "gator holes". The deeper water is particularly important in the dry season when fish and other aquatic organisms concentrate in these refuges. This provides the alligator with a readily available food supply and is attractive also to a wide range of fish-eating birds, such as heron and osprey.

During the breeding season, different species build different types of nests, but all are designed to keep the eggs warm until they hatch. Most crocodiles dig a hole in suitable ground and deposit the eggs in layers which are carefully covered with sand. The nest needs to be close to water to enable the female to splash water on to it to keep it cool.

Alligators build their nests above ground from leaves, branches and sediment. The material is shaped into mounds up to 2 metres (6 feet) in diameter and a metre (3 feet) high. Eggs are laid in the centre and as the organic material decomposes sufficient heat is generated to incubate them.

Small aquatic mammals

There are a number of small mammals that are specialized for life in and around water, and that show both physical and behavioural adaptations to their environment. One of the best examples is the two species of beaver, the American beaver (*Castor canadensis*) and the Eurasian beaver (*C. fiber*). These large rodents are excellent swimmers; to help propulsion they have webbed rear feet and a broad, flat tail, which they use like an oar. The thick fur of beavers provides both insulation and waterproofing. A beaver is also able to close its nose and ears when under water. The most notable behavioural adaptation of beavers is the building of dams to create the pools in which they build their lodges.

Otters, such as the Eurasian otter (*Lutra lutra*), have many of the same physical adaptations as beavers to life in water, such as webbed feet and waterproof fur. They have the added advantage of a slim body, which undulates easily to help swimming. Other small mammals that are well adapted to an aquatic life include the muskrat (*Ondatra zibethicus*), the fish-eating rat (*Ichthyomys stolzmanni*), the marsh mongoose (*Atilax paludinosu*) and the otter-civet (*Cynogale bennetti*).

Left The omnivorous Alaskan grizzly bear (*Ursus arctos*) has learned to take every advantage of its varied habitats. These bears have caught Pacific salmon that were making their way up the McNeil River to their spawning ground. The migration of the salmon and their subsequent death upriver provides an annual bonanza of protein for the bears.

Above **An American alligator** (*Alligator mississippiensis*) at rest. In this position, the mouth is flooded, but a muscular valve keeps water out of the throat. As long as the valve remains closed, the alligator can breathe freely through its nostrils. The two surviving species of alligator, American and Chinese, are the only crocodilians not confined to tropical regions.

Larger aquatic mammals

Perhaps the most famous large aquatic mammal is the hippopotamus (*Hippopotamus amphibius*). While the popular image of the hippo is one of an animal that seems to be permanently submerged up to its nose in water, this masks the hippo's vitally important role as a carrier of nutrients from land to water. Hippopotamuses graze on terrestrial vegetation, generally at night, but defecate in the water, where they remain for most of the day to keep cool. Their tails are used to propel and help disperse the nutrient-rich waste products into the water to the benefit of the aquatic animals.

Male hippos also spray dung to mark their land territories. Harvester termites (family Hodotermitidae) then collect particles of the dung to mark their own land territories. The termites collect the dung because it still contains partially digested grass fragments. The termites complete the breakdown of the grazed grasses and return the nutrients to the floodplain soils. In addition, fungus-growing termites (*Macrotermes* sp.) build mounds on the floodplains, and as the mounds are broken down by erosion they enrich the soil. This nutrient-rich soil is favoured by the shorter, more nutritious grasses, which are the preferred food of the harvester termites. Thus, the recycling of nutrients, involving hippos, termites and grasses, gradually enriches the floodplain soils.

Fleet-footed herbivores

The sitatunga (*Tragelaphus spekei*) is an elusive, shy, well-camouflaged antelope. It is the only large mammal to feed on and inhabit the papyrus swamp of central and East Africa. Living as far south as the Okavango Delta, it can move across the dense papyrus beds and soft terrain with some ease despite its relatively large size – males weigh over 100 kilograms (220 pounds). Long hooves, almost twice the length of those of similar-sized antelopes, spread out as it walks, providing good mobility in the wetland habitat. The sitatunga is further favoured by raised hindquarters, giving a slow, loping gait enabling it to move quietly around the thick reedbeds – an important advantage in avoiding predators. The sitatunga also avoids capture by submerging itself in water, with just its nostrils above the surface.

Sitatunga feed on grasses and other emergent wetland plants in addition to papyrus. This adaptability is important as annual movements as well as diet are

determined by the flood pattern. When the water is high, the antelopes move towards the dryland margins to feed on a variety of vegetation; when water levels drop the sitatunga move deeper into the swamp to survive almost exclusively on papyrus shoots. Such versatility is a key to survival.

Sitatunga raise their young on platforms of flattened vegetation, hidden in the extensive areas of swamp. Normally, however, well-trampled paths form a tell-tale network indicating favoured feeding areas and routeways. The main threat in the water is from crocodiles, but lions and leopards may attack on islands and young sitatunga are taken by pythons.

Antelope dominate the seasonally flooded grasslands of East Africa. One of the most abundant is the red lechwe (*Kobus leche leche*), sometimes described as a semi-aquatic antelope because of its particular ability to graze on the young shoots of sedges and other plants growing in shallow waters. Like the sitatunga, the red lechwe have elongated hooves for moving over soft ground. In fact, they can run faster through shallow water than on dry land. Their preference for the flooded environment stems from a combination of the availability of green vegetation and the capacity to escape from land predators such as lions, leopards, wild dogs and humans.

Fish

Rivers, lakes and coastal waters are home for an enormous diversity of fish species, most of which are dependent in some way on wetlands – whether for food, spawning, nursery or other habitat requirements.

Some species, such as the European carp (*Cyprinus carpio*), for example, migrate from rivers to seasonally inundated floodplain forest or grassland for spawning; others migrate from deepwater lakes to vegetated shallows; yet more move from the open sea to coastal mud flats, lagoons, marshes and mangroves.

In the case of migratory species such as the salmon (*Salmo salar*), the distances covered may be thousands of kilometres from ocean feeding grounds to the shallow gravel-bed "redds" (spawning grounds) of streams. Both American and European eels (*Anguilla rostrata* and *A. anguilla*) spend most of their lives in fresh or brackish water but return to the Sargasso Sea to spawn.

Distinctive groups of fish are adapted for life in different water conditions. The fish population of rivers often reflects speed of flow. The fast, or torrent, zone has clear, well-oxygenated, mineral-poor water, which flows over generally high, rocky terrain. This zone is inhabited by fish such as trout (*Salmo trutta*) and salmon (*S. salar*). Medium waters are characterized in northwest Europe by barbel (*Barbus barbus*), chub (*Leuciscus cephalus*), roach (*Rutilus rutilus*), rudd (*Scardinius erythrophthalmus*) and perch (*Perca fluvialis*).

In the intermediate zone between medium and slow-moving waters, pike (*Esox lucius*), carp, tench (*Tinca tinca*) and white bream (genus *Abramis*) may be more typical. Where there is little or no gradient, the current is tidal and the water brackish to saline, flounder (*Platichthys flesus*) and mullet (for example *Crenimugil labrosus*) are usually widespread. They are able to tolerate seawater as well as fresh water during times of flood and at low tide. Migratory species such as salmon, trout and eel can survive all levels of salinity and are commonly described as *euryhaline*.

Left **The hippopotamus has an almost completely aquatic life style and has become superbly adapted to life in rivers. The nostrils are positioned on the upper surface of the snout and protrude so that the hippo can float with almost all of its bulk below the surface.**

Bottom left **Waterbuck (*Kobus ellipsiprymnus*) and sacred ibis (*Threskiornis aetheopicus*) feeding in East Africa. Where possible waterbucks graze only on aquatic plants, while the sacred ibis will take fish and amphibians.**

Below **Rainbow trout (*Salmo gairderi*) gather in a New Zealand stream. Trout were introduced to the country as a sport fish, and are now well established. Classifying salmon and trout is extremely difficult. All species migrate to some degree, and although most trout are confined to rivers throughout their lives, some rainbows continue to migrate out to sea like their salmon ancestors. Because of their migratory habits, some species are able to adapt to water at all levels of salinity between fresh water and sea water.**

High and dry

Of the several species of lungfish, the African lungfish (*Protopterus aethiopicus*) is the largest. It grows to about 2 metres (6.5 feet) long and lives in the rivers and lakes of East and central Africa. Other species of lungfish are found in Australia and South America. As their name suggests, lungfish have lunglike breathing organs with which they can breathe air to supplement the usually poor oxygen supply they receive from their swampy river and lake habitats. In addition, these fish live in regions that experience seasonal periods of flooding and desiccation. When the water level becomes dangerously low, the lungfish digs a burrow, covers itself in a slimy mucus secretion to prevent its skin from drying out and breathes air. The fish will then slow its metabolic rate to conserve energy until the water rises again. This form of inactivity is known as *aestivation*.

Shorebirds, ducks and geese

"Shorebirds" (or "waders") is the name given to a group of 214 species of small to medium-sized bird. Shorebirds prefer shallow water and wetland habitats during the non-breeding season, and are usually gregarious and often highly migratory. The majority of the 214 species of shorebird belong to one of two families – the Charadriidae, which is made up of the lapwings and plovers, and the Scolopacidae, which comprises the sandpipers and snipes. Particular adaptations have evolved within the shorebird group that enables the various species of bird to live and breed together successfully in their wetland environment.

Beaks and feet

Beaks of the shorebird group are highly adapted and sufficiently versatile to obtain food of various types in highly efficient ways, often exploiting

While gulls and terns exploit only the top few centimetres of water for feeding, cormorants often dive more than 40 metres (130 feet) underwater to reach their prey.

Above **The darter (*Anhinga anhinga*) fishes quiet inland waters. Darters swim with their bodies below the water and only their heads and necks above the surface.**

Above right **The beaks of waterbirds are adapted to their feeding habits. The roseate spoonbill [A] (*Ajaia ajaia*) rakes through bottom ooze to trap invertebrates. The American avocet [B] (*Recurvirostra americana*) has an upturned beak which it uses in a sideways motion to skim vegetation from muddy water. The painted snipe [C] (*Rostratula benghalensis*) uses its beak to pluck worms, insects and seeds from marshy ground and lake shores. The scarlet ibis [D] (*Eudocimus ruber*) probes deep into coastal mud flats and mangrove thickets in search of crabs and molluscs.**

Right **Lungfish embedded in mud during aestivation. The layer of mucus that covers the body provides a substitute for the normal aquatic habitat.**

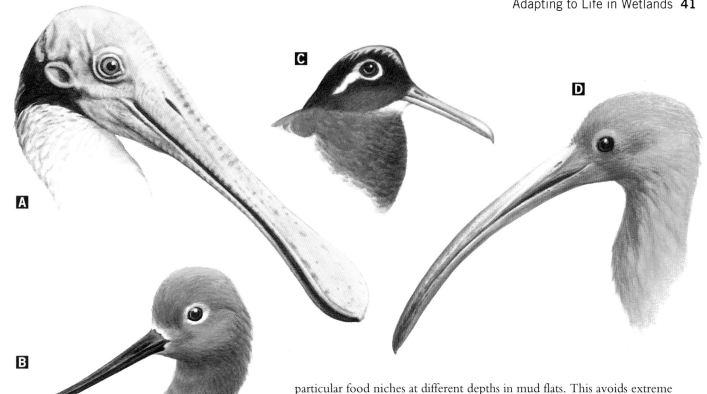

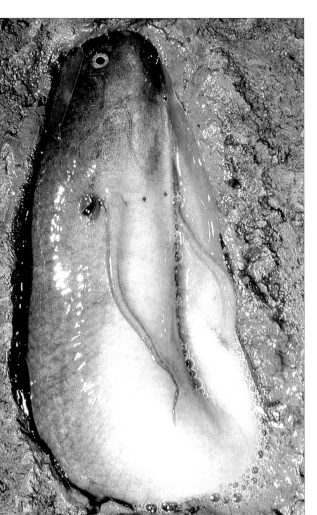

particular food niches at different depths in mud flats. This avoids extreme competition for food and enables very high numbers of a variety of bird species to feed in the same physical space.

Curlews have a long, strong curved bill which can be probed deeply into the mud for food. Snipe do the same, but their beaks are long, straight and slender, enabling more sensitive deep-probing of soft sediment. The avocet is one of the few birds with an upturned bill which enables it to explore the upper layers of the mud.

Specialized feet are necessary for many birds to survive in their wetland environments. Greenshanks, redshanks and most other waders have long toes, spread wide apart to increase the surface area of the foot. This allows the birds to move freely across soft mud and other sediment, where other birds and animals might become bogged down.

Other species of waterbird have developed legs for swimming and diving. Those of ducks and geese are near the rear of the body, where they can be used more effectively as paddles. The disadvantage of this arrangement for movement on land is only too apparent. The inconvenient distance between the legs and the bird's centre of gravity results in the familiar ungainly "waddle". In the case of the osprey the scales on the footpads are armed with small spines which help the bird to secure and carry slippery fish.

Diving techniques

While gulls and gannets can use the momentum of a sky dive to overcome the resistance of water, birds that have to dive from the surface must rely on the rear leg positioning, which is so marked in grebes and diving ducks, to give unimpeded thrust. Underwater swimmers are usually heavier than land birds of a similar size. Additional reduction in buoyancy can be obtained by compressing feathers or air sacs, forcing out the air. The dabchick uses this technique to float at different levels in the water. The dipper, however, can walk along the bottom of a stream by facing the flow, tilting up its back and allowing the force of water to hold it down. Its paddle-shaped toes and webbed feet help the dipper to walk through water.

The webbed toes of ducks, geese and divers present an enlarged surface area for powerful propulsion across the surface or underwater. Some sea-going ducks, such as scoters and other birds, including cormorants, use not only their legs but also their wings for swimming. Different species of diver tend to utilize different depths, so that like the waders, which have different bill lengths and structures, not all have the same underwater niche. While gulls and terns exploit the first top few centimetres of water, cormorants dive more than 40 metres (130 feet) to reach their prey.

Even a Florida sunset cannot detract from this desolate scene in the Everglades. Saltwater encroaching from the sea has killed off these bald cypresses. Formerly, there was a positive flow of freshwater southwards through the Everglades, which prevented contamination by seawater. The diversion and use of inland freshwater resources has diminished that flow to a point where it can no longer hold back the sea.

Wetland Loss

When Christopher Columbus set sail from southern Spain in August 1492, he epitomized the spirit of a continent on the threshold of the modern era. In the years that followed, mariners and explorers visited every corner of the globe searching for land and riches to serve the developing states of Europe. Simultaneously, the same creativity and enterprise in agriculture and industry drove the agricultural and industrial revolutions which swept across the Continent.

The agricultural and industrial revolutions not only changed the social, economic, and political face of Europe and ultimately the world, but they also set into motion a process of ecological change and devastation which continues today. In the words of one historian, the Columbian achievement "enabled humanity to achieve, . . . the transformation of nature with unprecedented proficiency and thoroughness, . . . altering the products and processes of the environment, modifying systems of soils and water and air". Among the systems that received special attention from this thriving European culture, both at home and in the newly discovered and settled lands, were the marshes, floodplains and other wetlands, most of which were perceived to be disease infested and obstacles to development. Once drained, these could be put to productive agricultural use.

Working with water

Europe during the 1400s was, of course, not the first civilization to modify natural wetlands. Presently, almost 70 per cent of the world's population lives on sea coasts, and in many regions, river valleys and lake shores have been settled for thousands of years. Over much of Asia, a diversity of cultures and empires had been built upon the control and exploitation of the regions' wetland systems. The civilizations of the Indus Valley and Angkor in Indochina drew much of their economic strength and stability from their efficient manipulation of the Indus and Mekong rivers.

In studying the use of water in Asian society, historians have made clear distinctions between hydraulic civilizations, which controlled water flow through dikes and dams, and aquatic civilizations, which exploited the annual cycle of river flood and adjusted to its excesses by, for example, building their houses on stilts. In general, the hydraulic form of life developed inland, where water was more seasonal and needed to be controlled in order to be brought most efficiently to the best agricultural land. In contrast, the aquatic civilizations inhabited the deltas and floodplains where water was abundant as it moved to the sea.

The concept of working with nature was largely absent from Columbian Europe. Instead, the hydraulic culture which sought to control and dominate the aquatic environment governed much of human society's relationship with wetlands in the subsequent 500 years. Along the coast of Europe, the Dutch began to dike their shallow sea and turn its bed into reclaimed agricultural land, known as polders. In Britain and France, similar investments were made in channelling the major rivers. And as the new colonies became established in the Americas, Africa and Asia, this hydraulic technology was exported, and indeed continues today.

Below left **Salt marsh filled with sand to provide land for new housing in North Carolina, USA.** Increased suburbanization in the United States has placed additional pressure on wetlands that require relatively little effort to prepare them for housing.

Right **Trees killed by salinity on the floodplain of the Murray River, Australia.** The waters of the Murray–Darling Basin have a high salinity, but the problem is aggravated by irrigation. Water is diverted from the river system and from artesian sources. Saline waste water from irrigated farmland is often drained on to the flood plain, with the result that salt levels are increasing.

Bottom **Vientiane in Laos is a good example of an aquatic civilization.** By building their houses on stilts, the people are working with the hydrological system, rather than trying to control it with dikes and dams.

Agricultural devastation

The loss of wetlands worldwide, which some specialists estimate as being in the order of 50 per cent of those that once existed, is the direct consequence of this hydraulic vision of the world. And although detailed figures are difficult to obtain, the limited information tells a sorry tale. In the United States alone, some 54 per cent – 870,000 square kilometres (360,000 square miles) – of the wetlands which once existed are believed to have been lost, while in some states the proportion lost is even higher. Taking the nation as a whole, 80 per cent of wetland loss has been to agriculture. For 200 years, the conversion and destruction of wetlands was actively encouraged by the United States federal government. In 1763, George Washington set up a company to drain the Great Dismal Swamp of Virginia and North Carolina, and convert it to agricultural land. Although that particular scheme failed, the nation's attitude towards wetlands was set. Further acts in the mid-1800s, known as the Swamp Land Acts, underlined the conversion attitude. States were encouraged to build levees, carry out drainage projects and destroy the mosquito-infested wetlands. Following these acts alone, 260,000 square kilometres (100,000 square miles) of wetland areas were lost.

In Europe, the rates of loss are less well documented, but, with the continent's high population density and longer history of economic development, the conversion of natural ecosystems is believed to be greater. Forty per cent of the coastal wetlands of Brittany have been lost since 1960, and two-thirds of the remainder are seriously affected by drainage and similar activities. In southwest France, some 80 per cent of the marshes of the Landes have been drained. In Portugal, some 70 per cent of the wetlands of the

western Algarve, including 60 per cent of estuarine habitats, have been converted for agricultural and industrial development. In New Zealand, it is estimated that over 90 per cent of wetlands have been destroyed since European settlement, and drainage continues today. Fourteen per cent of the remaining freshwater wetlands were drained in the North Island in the five years between 1979 and 1983. For the developing world little detailed information is available on rates of wetland loss. However, what is available has given rise to concern that entire ecosystems are now under threat. In the Philippines, some 3,000 square kilometres (1,000 square miles) – 67 per cent – of the country's mangrove resources were lost in the 60 years from 1920 to 1980, some 1,700 square kilometres (650 square miles) being converted to ponds for farming shrimp and milkfish. In Nigeria, the floodplain of the Hadejia River has been reduced by more than 300 kilometres (190 miles) as a result of dam construction; and in Brazil most estuarine wetlands have been degraded as a result of pollution.

It has been estimated that salinization, waterlogging and alkalization of soils affect some 400,000 square kilometres (150,000 square miles) – 20 per cent of the world's irrigation schemes.

The consequences of wetland loss

As we have seen, many wetland regions have been destroyed because society views their destruction as either good in itself, or as a small price to pay for the benefits expected from wetland conversion. Today, such policies are increasingly condemned as short-sighted, as well as being socially and economically indefensible.

Dams, of which there are 114 major projects planned in West Africa alone, and other river basin schemes have come under special criticism. Many have destroyed wetlands while falling far short of their predicted benefits. The result has been hardship for those populations dependent upon the floodplain and other wetlands downstream. In Nigeria, fish catches and floodplain harvest declined by over 50 per cent in an area extending 200 kilometres (120 miles) downstream of Kainji Dam, and losses in yam production of some 1,000 tonnes were reported in the lower Anambra Basin of eastern Nigeria. In the arid north of the country, the loss of crop production on the Sokoto River floodplain following completion of the Bakolori Dam was estimated as US$7 million at 1974 prices. Farmers had to leave home to find work in the cities or on other farms as migrant labourers to support their families. In northern Cameroon, construction of the Semry II irrigation project on the Logone River has reduced flooding downstream, causing a collapse in the fish yields and making it impossible to grow floating rice, the principal cereal crop of the Kotoko community.

The issue of floodplain development highlights the general principle that wetlands will be destroyed where people envisage putting their water to a more productive use. But just how valid is this assumption? When efficiency is measured in terms of profit per unit of water, data from African floodplains suggest that over time there is very little difference between traditional extensive methods of agriculture and intensive rice cultivation. And when the costs of the capital investment are taken into account rice cultivation can actually lose money.

A question of economics

Some of the products and services of wetlands are sold, such as commercial fisheries, meat and skins from grazing herds and crops. But many wetland values do not have identifiable markets, such as water purification and flood protection. Because these values are "free goods" they tend to be ignored in the economic calculations that decide whether wetlands should be conserved or developed. The result unfortunately usually favours development and, with it, wetland degradation. Private landowners, for example, frequently decide to drain their wetlands because they expect to earn more from growing crops

than from leaving them in their natural condition. They may be perfectly aware of the role wetlands play in groundwater recharge and discharge, flood control, fisheries support and nutrient retention, but these public benefits count for less than their private profit.

To compound the problem, many governments which on the surface actively encourage wetland conservation, are, by some of their policies, causing the opposite to happen. Wetland drainage in Europe, for example, where incidentally all countries with the exception of Luxembourg are parties to the Ramsar Convention, was stimulated by the artificially high prices paid for a number of crops under the EC's Common Agricultural Policy (CAP). Under the CAP, intervention prices were at times well above market prices. Wheat, for example, was 40 to 60 per cent above world market prices during 1978 to 1980. As a result the conversion of lowland wet grazing meadows for growing winter wheat became a very profitable financial investment for individual farmers.

Irrigated land, much of which was formerly natural wetland, is also meeting serious problems in many areas of the world. It has been estimated that salinization, waterlogging and alkalization of soils affect some 400,000 square kilometres (150,000 square miles) − 20 per cent of the world's irrigation schemes. As a result, hundreds of thousands of square kilometres of irrigated land, and the revenue from the crops, are being lost each year. In other instances, irrigation was achieved at the cost of disrupting normal water supply and consequently has compromised the

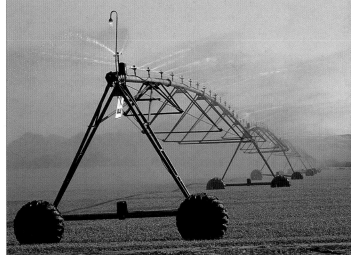

Top **Spruce forest killed by the effects of acid rain in Krkonose National Park, on the Polish/ Czech border.** Produced in the atmosphere from industrial pollution, acid rain has a double deadly effect. Falling as precipitation, the acid chemicals attack the exposed surfaces of vegetation, destroying foliage. Acid rain run-off is likely then to enter the groundwater, affecting the hydrology of many wetland habitats. In particular, it collects in ponds and lakes, killing off the flora and fauna.

Above **Centre-pivot irrigation systems use both surface- and groundwater.** Spraying is a wasteful method of irrigation as losses through evaporation are very high. In many regions, aquifers are being depleted for irrigation, and the hydrology of entire regions is threatened.

Left **Tin-mine in Thailand.** The mining and extraction of metal ores and minerals involves huge quantities of water. Even with filtration systems it is impossible to prevent toxic contaminants entering the environment.

The Causes of Wetland Loss

	Estuaries	Open coasts	Flood-plains	Freshwater marshes	Lakes	Peatlands	Swamp forest
Human Actions							
Direct							
Drainage for agriculture and forestry; mosquito control	■	■	■	■	▦	■	■
Dredging and stream channelization for navigation; flood protection	■	□	▦	□	□	□	□
Filling for solid waste disposal; roads; commercial, residential and industrial development	■	■	■	■	■	▦	□
Conversion for aquaculture/mariculture	■	□	□	□	□	□	□
Construction of dikes, dams and levees; seawalls for flood control, water supply, irrigation and storm protection	■	▦	▦	▦	▦	□	□
Discharges of pesticides, herbicides and nutrients from domestic sewage; agricultural runoff; sediment	■	■	■	■	■	■	□
Mining of wetland soils for peat, coal, gravel, phosphate and other materials	▦	▦	▦	□	■	■	■
Groundwater abstraction	□	□	▦	■	□	□	□
Indirect							
Sediment diversion by dams, deep channels and other structures	■	■	■	■	□	□	□
Hydrological alterations by canals, roads and other structures	■	■	■	■	■	■	□
Subsidence due to extraction of ground-water, oil, gas and other minerals	■	▦	■	□	□	□	□
Natural Causes							
Subsidence	▦	▦	□	□	▦	▦	▦
Sea-level rise	■	■	□	□	□	□	■
Drought	■	■	■	■	■	▦	▦
Hurricanes and other storms	■	■	□	□	□	▦	□
Erosion	■	▦	□	□	□	▦	□
Biotic effects	▦	□	■	■	■	□	□

□ Absent or exceptional

▦ Present but not a major cause of loss

■ Common and important cause of wetland degradation and loss

Left **"Protected" wetland in Louisiana, USA.** The uninviting appearance of many wetlands causes people to consider them "natural" dumping grounds for domestic waste, unwanted vehicles and construction materials, etc. Once dumping starts, it is very difficult to stop people continuing to dump.

Below **Glen Canyon Dam, Arizona, USA.** The dam is typical of the many hydro-engineering projects that have been undertaken during the last 70 years. Such schemes usually have wholly admirable objectives, such as flood control or the provision of electricity. However, while such projects are usually judged a success in terms of their primary aim (in the case of Glen Canyon to provide irrigation water), it is now clear that planners often paid scant attention to the potential for collateral damage. A large dam can radically alter the pattern of flooding in a river system, resulting in disastrous consequences for downstream wetlands.

Far left **An extensive area of mangroves is almost completely destroyed, cut for firewood in the Sengal Delta, Mauritania.** In many parts of the developing world, fuelwood is the only energy source for domestic lighting and cooking.

long-term viability of the investment. In Peninsular Malaysia, 90 per cent of freshwater swamps have been drained for rice cultivation. However, without the freshwater normally supplied by the swamps, rice production has been far below expectations.

In the past, loss of these wetland benefits generally has gone unremarked because the relatively strong national and household economies of industrialized nations could afford to pay for the consequences. Flood control and water purification, once provided free by wetlands, are today maintained by dams, dikes and other measures that were financed by increased taxes.

The increasing cost of wetland loss to the industrialized world, however, has reached such proportions that major efforts are now being made to conserve the remaining wetlands as functional economic units. For example, in the United States, where the loss of privately owned wetlands has resulted in major public costs, federal aid for drainage activities has now been removed, and crop subsidies are no longer available for land-owners who drain their wetland areas.

Real dependants

In the developing world, the rural economy and human well-being are even more closely dependent upon the wetland resource. Only rarely are national or household economies strong enough to replace goods or services once provided free by wetlands. The consequences of wetland loss are, therefore, fundamentally more severe in developing countries. There, loss of wetland resources leads not only to increased taxes, but to flood damage, contaminated water, human suffering and death.

Similarly, in societies that rely on wetlands for fish protein, pasture, agricultural products or timber, any reduction in productivity is felt acutely. At best an increased proportion of the household budget has to be spent on subsistence and less on housing or education, while in many cases the reality is a lower-quality diet, or even a decline in total food intake. In the more extreme cases, as in many African floodplain systems, it can lead to rising mortality and emigration.

Rural Development

In April 1989, a dispute over floodplain pastures in the Senegal Valley triggered ethnic violence which left over 1,000 people dead and tens of thousands homeless. While the ultimate reasons for this violence have a long and complex history, the immediate cause, a dispute over a valuable wetland resource, dramatically underlines the link between effective management of wetlands and human well-being. If well managed, wetlands can help to meet the needs of a rising population; while their degradation and loss, however, can worsen the already intense pressures upon rural communities in many parts of the world.

The traditional image of wetlands is one of an inaccessible waterlogged swamp that harbours disease-carrying mosquitoes. But times have changed. As more and more wetlands have been lost and others severely degraded we have grown to appreciate their uses. The challenge today is to develop approaches to wetland management which can ensure that these benefits are made available on a long-term basis. Wetlands are productive places and they can play a central role in creating sustainable development.

Today's strategies

In rising to this challenge, modern society has much to learn from the many systems of sustainable wetland use which have been practised for centuries by rural communities in Africa, Asia and Latin America. For example, over much of arid Africa, migratory pastoral communities have tuned their annual cycle to the river flood, moving out onto arid, rain-fed rangelands as the waters rise, and back onto the floodplain to graze the rich pastures as the waters recede. In the larger floodplain systems, such as the Inner Niger Delta in Mali, fisherman follow similar cycles, moving within the river system to fish in areas of highest fish abundance at specific times of year. Even immobile communities adjust their lives to the river flood, planting rice as the river rises and harvesting it as it falls.

However, these traditional practices need to be adjusted to today's conditions. Many traditional systems of grazing, fishing and wetland agriculture have been developed under tightly regulated conditions, which recently have begun to break down. Rising population and associated increase in demand for agricultural land, for pasture and fishing, are today straining, and in many cases breaking, these traditional methods.

At the same time, drought and man-made changes in river flow have decreased the capacity of wetlands to absorb existing pressure. In many countries, these factors have been exacerbated by placing management control in the hands of central government, rather than in local institutions, which often have much greater knowledge of the resources in question and their management needs.

As a result of these changes, wetlands are no longer capable of supporting the demands placed upon them. In response, many governments have chosen to invest billions of dollars in converting natural wetlands for intensive agriculture because this is perceived as the most effective means of increasing food production. However, this decision is now questioned openly, and many

Top **Street scene in Malaysia. A drain feeds a stream that discharges into the Malacca Straits. Raw human sewage and household chemicals pass untreated into the** environment where they contribute to the pollution of coastal waters and wetlands. Urbanization can place tremendous pressure on natural waterways in towns and cities.

"Pressure on resources increases when people lack alternatives. Development policies must widen people's options for earning a sustainable livelihood, particularly for resource-poor households and in areas under ecological stress." (World Commission on Environment and Development).

Below *Anopheles* mosquito on human skin. Mosquitoes, which breed prolifically in wetland areas, are the vectors (carriers) for the potentially fatal disease malaria.

In the recent past, large areas of wetland have been drained in an attempt to eradicate the disease, particularly in parts of Europe and the United States.

intensive agricultural development schemes are criticized increasingly as being at best inappropriate and expensive, while in actual fact many are widely acknowledged to have had a substantial negative impact.

Changing attitudes

Although understanding of the value of wetlands and associated investment in wetland conservation has perhaps grown most rapidly in the United States, similar changes in people's perception of wetlands have given rise to a range of conservation initiatives all over the world. In Uganda, for example, the Ministry of Environment Protection is finalizing a national wetland policy that is in direct response to rising local concern over environmental and social consequences of wetland loss in the southwest of the country. In Sri Lanka, concern for the country's coastal wetlands has encouraged the Natural Resources, Energy and Science Authority (NARESA) and Department of Forestry to collaborate with other governmental and non-governmental institutions in preparing a management plan for the country's mangroves.

On a regional scale, the southern African Development Coordination Conference (SADCC) is preparing a programme for conservation and management of wetlands in Southern Africa, while in Southeast Asia, the Mekong Secretariat

Above **Shanabala nomads at a waterhole near El-Obeib, Sudan. In the arid Sahel, the nomads water their animals whenever possible. Over the next few years, the demands upon** water and wetlands will rise substantially as populations increase and global climate change is likely to cause higher temperatures and reduced rainfall in arid regions.

is preparing a similar initiative for the wetlands of the Mekong Basin. On a global scale, a range of international organizations are expanding the wetland activities, including IUCN – The World Conservation Union and the World Wide Fund for Nature (WWF). The Ramsar Convention (see page 12) is flourishing with the number of Contracting Parties having increased from 23 at the beginning of 1980 to 55 at the beginning of 1990.

Northern landscape – the Mackenzie Lowland in Canada's Yukon Territory. Dense, coniferous forest defines the contour line marking the boundary between wet and dry land. The trees are confined to the low ridges and patches of higher ground that rise above a landscape of countless small lakes and permanent marsh. The effects of periodic flooding are shown by the areas of unforested land around some of the lake margins.

Atlas of the World's Wetlands

Unlike many other endangered habitats, wetlands are not restricted to specific regions of the world. Their occurrence does not rely on direct rainfall, temperature or altitude, and for this reason examples of wetland systems are found in some of the hottest places on Earth as well as some of the coolest, from the breathtaking heights of the Andean Altiplano to Australia's Lake Eyre, which lies below sea-level. Wetlands are found everywhere, and a great many may soon be lost forever.

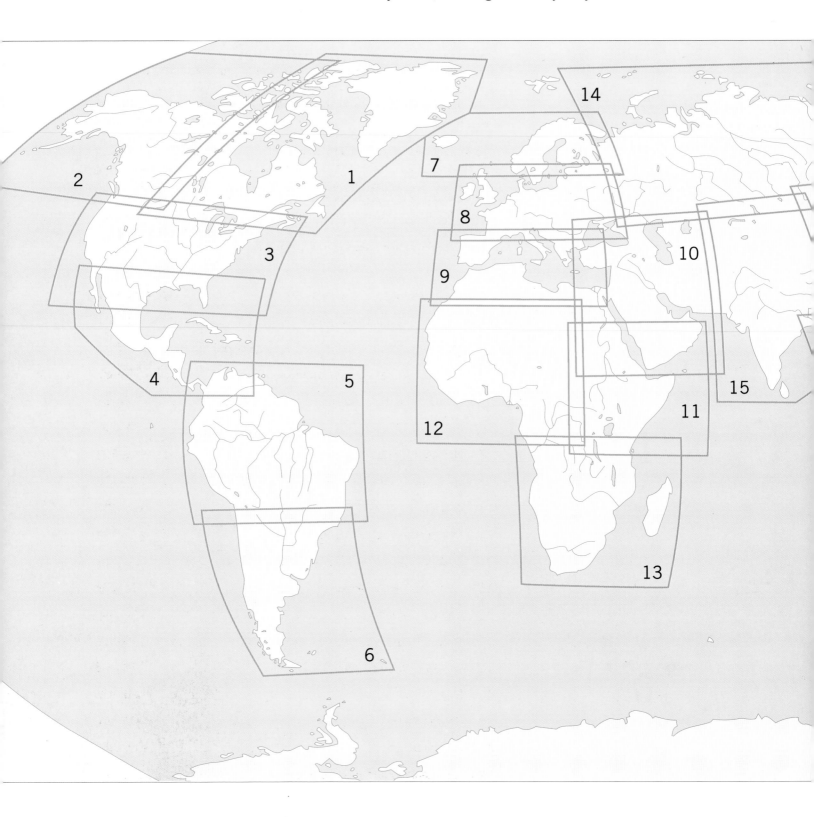

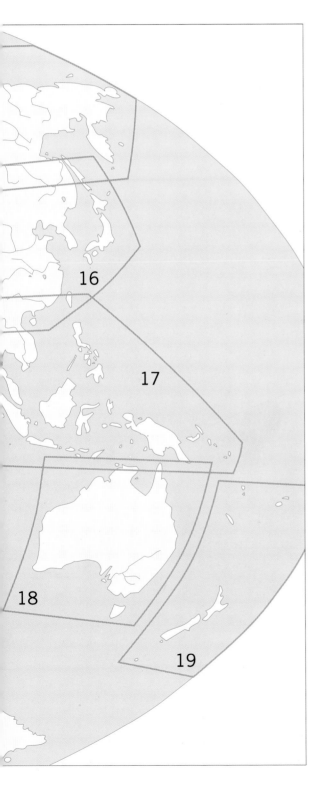

The many and varied wetlands of the world have ensured that the Atlas section of **Wetlands in Danger** is the largest in the book. Yet this section is not simply a cartographical exercise that provides information on the geographical location and type of wetland found throughout the world.

Each of the 19 map sections (listed below) begins with an overview of the region's wetlands, informing of their extent and type, before going on to focus on specific wetland systems that are either of special value to the region's population or of interest in terms of their biodiversity. Most of the areas discussed within the Atlas section are under threat in one way or another, whether by agricultural or urban conversion or by an alteration of their associated hydrological signatures. The often complex socio-political reasons for many of these regions' wetland degradation are clearly explained, and information is given of any conservation policies that have been set up to stop the threat of conversion or to reverse its affect.

The Making of the Maps

The maps contained in this atlas represent the most comprehensive and accurate assessment of the world's wetlands compiled to date. The number of sources varies form continent to continent, but are broadly based on a series of regional wetland directories that have been prepared by IUCN – The World Conservation Union and partners, including The World Wide Fund for Nature (WWF), The United Nations Environment Programme (UNEP) and The International Waterfowl and Wetlands Research Bureau (IWRB). So far, directories have been published for Latin America, Asia and Africa, while one for Oceania was, at the time of publishing, in the final stages of preparation and proposals are under development for the Middle East.

In preparing this work, The World Conservation Monitoring Centre (WCMC) has, for each region, sort to map the broad grouping of wetland ecosystems as defined by the Ramsar Convention (see page 12). Yet one of the major messages of this book is that although wetlands are a common feature of the natural landscape worldwide, the dominant types of wetland vary from region to region. For this reason, and in recognition of the limited data available for some regions, WCMC has exercised necessary flexibility in representing this diversity in each region. For example, in Africa there is a simple "swamp forest" category, while in South America the corresponding wetland type is combined as "swamp forest/flooded forest". This reflects the differing ecological conditions as there is no widespread equivalent in Africa of a forest which is flooded to a considerable depth only on a seasonal basis.

From data to maps

For southern Asia and Africa, the authors of the directories were asked to identify on 1:1 million scale topographic maps the sites referred to in the directories. These were then digitized at WCMC and the outlines stored in a Geographic Information System (GIS). The areas shown were then classified into a number of categories which correspond with the colours used in this atlas. The GIS allows the information to be presented in a variety of scales and projections to suit the requirements of the map. Although the continent of Africa is represented in four maps for this atlas, it would be equally possible to show the wetlands of a single country, such as Benin, at a much larger scale. This sets the maps apart from any earlier compilations of wetland distribution and explains the fine detail which is apparent at the scales shown here. Information for Oceania and the Middle East has been compiled on a similar basis by consultation with the authors of the directories in preparation.

North America

For North America, we are indebted to the United States Fish and Wildlife Service (USFWS), who have been compiling an inventory of the wetlands of the United States (see page 71), and who have published maps of the wetlands of Alaska and the conterminous 48 states; these were digitized in a generalized form at WCMC. USFWS, together with another urban and ecological development organization, Secretario de Desarrollo Urban y Ecologia (SEDUE) and Conservation International have recently compiled a map of the wetlands of Mexico and have been kind enough to make available a copy in digital form for use in this atlas. For most of the rest of Central America, the IUCN wetlands programme has prepared wetland maps which were digitized specifically for this atlas.

Northern Asia and Australia

For the northern Asia map, a map of nature conservation of the former USSR was used. The map was prepared by Moscow State University and from this, types of vegetation associated with wetland habitats were digitized.

A map of Present Vegetation, produced by the Australian Surveys and Land Information Group was used for Australia, with the category for mangrove being overlaid with additional data from topographic maps.

1 The first stage in producing an up-to-date map of wetland systems is to locate on an existing outline map basic information such as lakes, rivers and areas of relevant vegetation (such as mangroves). These are read off the outline map with the use of an electronic pen or digitizer, which transmits coordinates to a computer.

2 When all the information required has been transferred on to the computer, a printed copy of the map is produced.

It is at this stage that the cartographer can get an impression of how the final map will look. Adjustment to the colour-coding and types of symbol used can be made.

3 Careful checking of the finished map against all the sources from which the map was compiled is required to ensure that the information can be interpreted not just by scientists and geographers, but by politicians, conservation policy makers and concerned individuals.

Right Women carrying fishing baskets to a wetland site in Botswana. The mapping of existing wetland ecosystems is perhaps most important for developing countries. The developed Western world has already lost the vast majority of its wetlands, the cost of which has been offset by industry and manufacture. But for the developing nations, the conservation of their "life-supporting" wetlands is essential. Through meticulously mapping these systems, it is possible to identify those that are under most pressure, recognize the causes and then, through reasoned discussion and dialogue, set in place programmes of conservation that will be of benefit to local communities, not just in the short term, but for decades to come.

European sources
Ironically, it is for Europe that it has been most difficult to compile a summary map. There is no comprehensive wetlands directory for this region, and the other information available through such initiatives as the Coordination of Information on the Environment (CORINE) programme of the European Commission, is so detailed that it was not possible to use it to compile summary maps of the area. Furthermore, the CORINE programme was organized on a country by country basis and the quality and quantity of information varies from extremely detailed in some EC countries to zero for countries outside the community. The wetland sites for Europe are therefore best illustrated by the extensive network of Ramsar Sites which follow the major wetland features.

Eastern Canada and Greenland

Eastern Canada is home to a diversity of wetland types which cover more than 570,000 square kilometres (220,000 square miles) – an area the size of France. The wetlands range from arctic tundra ponds to peatlands and temperate fens; from freshwater swamps to coastal marshes.

In some parts of eastern Canada, wetlands have come under severe pressure as a result of human settlement. In southern Ontario, since major settlement began in the mid-1800s, over 68 per cent of the wetlands have been converted to other land uses, either for agriculture or urban expansion. The wetlands around the Great Lakes have, unfortunately, been among the most severely affected.

However, 70 per cent of wetlands in eastern Canada are forested peatlands, lying in the boreal (more northerly) forest region. Here winters tend to be long and harsh, and summers short, and the peatlands have experienced little development pressure. The main reason for development control lies in the hands of the Canadian forestry industries. Their management of the wetlands encourages sustainable forestry practices. These include sensitive use of wood products. However, recent proposals for major expansion of hydroelectric facilities in the boreal and subarctic regions of Quebec, Ontario and Manitoba are threatening a range of wetland and upland habitats, particularly at coastal and estuarine sites. Large-scale flooding of several areas, and alteration in the natural flow patterns and water levels in many rivers and estuaries, is expected as a consequence.

Conservation policies

In the face of these pressures, the governments of several eastern Canadian provinces are implementing wetland conservation policies. In 1989, the province of Prince Edward Island became the first Canadian jurisdiction to put in place wetland protection regulations. Since then, Quebec has introduced measures to protect wetlands along major lakes and waterways such as the St Lawrence River and its estuary, the areas under greatest threat. In New Brunswick, potential conversion projects of all wetland sites over 2 hectares (5 acres) must undergo strict environmental assessment regulations. The province of Ontario is presently consulting public bodies on the question of a wetland management policy which will be enacted through provincial planning legislation. Under this scheme a number of sites will be evaluated and protection extended to the most important wetlands. In addition, the Canadian Government has announced "The Federal Policy on Wetland Conservation", an initiative that focuses on wetlands on federal lands and in areas of federal jurisdiction. Canada is one of the first nations in the world to establish such a policy.

Protected sites

A total of 16 wetland sites designated of international importance by the Ramsar Convention are located in all six of the provinces of eastern Canada and the eastern portion of the North West Territories. These include sites such as the Polar Bear Pass National Wildlife Area and the Dewey Soper Migratory Bird Sanctuary on the Great Plain of the Koukdjuak in the North West Territories, and the Polar Bear Provincial Park and several bird sanctuaries in the Hudson-James Bay area. Other Ramsar Sites include key wetlands along the St Lawrence.

Preserved wetlands in the lowlands or coastal areas found on Canada's Atlantic coast include the popular Cap Tourmente National Wildlife Area near Quebec, and the Long Point National Wildlife Area and Point Pelee National Park in southwest Ontario. The Western Hemisphere Shorebird Reserves at Shepody Bay and Mary's Point in New Brunswick, which are famous for the hundreds of thousands of overwintering shorebirds, are also important sites.

Coastal and estuarine salt marshes are the wetlands facing greatest threat from development along Canada's Atlantic coast. These are critical areas for migratory shorebirds, waterfowl and a major portion of all commercially harvested fish and crustaceans which depend upon the wetlands for part of their life cycle. Since settlement, about 65 per cent of salt marshes in New Brunswick, Nova Scotia and Prince Edward Island have been altered or destroyed by the development of ports and harbours, or diking to permit agricultural expansion.

Greenland

The Greenland Ice Cap covers 80 per cent of the island's landmass, with only parts of the coastal strip in the northeast and west remaining free in summer. This narrow strip is mainly mountainous, so the area of wetlands is small, and the deltas and estuaries are almost lifeless. The most important freshwater wetlands are usually a mixture of small lakes, marshes and moist tundra. In the Low Arctic zone, willow (*Salix* sp.) shrub is an important component. These inland wetlands provide the breeding habitat for six species of geese, one of which, the Greenland white-fronted goose (*Anser albifrons*), is an endemic subspecies. The High Arctic wetlands are also moulting areas for pink-footed geese (*Anser brachyrhynchus*) and barnacle geese (*Branta leucopsis*), and nesting sites for large numbers of shorebirds, such as dunlin (*Calidris alpina*), sanderling (*C. alba*), knot (*C. canutus*) and ringed plover (*Charadrius hiaticula*).

1 Jameson Land

Jameson Land, a peninsula found on the east coast of Greenland, has one of the largest lowland tundra areas in the country. Some 1,250 sq km (480 sq miles) of the tundra region has been designated a Ramsar Site. Several rivers cut through the area and there are salt marshes on the coast. The many species of breeding bird found in the region include red-throated diver (*Gavia stellata*), barnacle goose, long-tailed skua (*Stercorarius longicauda*) and whimbrel (*Numenius phaeopus*). The largest number of birds, however, are non-breeding geese, and 5,000 pink-footed and 2,500 barnacle geese come here to moult. Bulky, long-haired musk oxen (*Ovibos moschatus*) are abundant in this region and the scores of moulting geese are prey to the numerous arctic foxes (*Alopex lagopus*).

Ramsar Sites
Parks and Reserves
Water Bodies
25-50% Wetland
50-100% Wetland
Prairie Pothole Region
Important Bird Areas

The Great Lakes Wetlands Action Plan

Wetlands along the shores of the Great Lakes, particularly Lakes Erie and Ontario, are among the most threatened wetlands of both Canada and the United States. Studies of the shorelines of Lake Ontario since the early 1800s show that 90 to 95 per cent of wetlands in this region have been destroyed to serve other land use needs.

Many shoreline swamps and marshes, and river delta marshes, have been developed into ports and harbours, industrial areas, diked for agriculture, or buried under other forms of urban development. Remaining wetland sites have been severely affected by toxic contaminants, artificial barriers against the variable seasonal water levels, and by storm effects. Pressure to develop shoreline wetland sites for building holiday homes also remains high. Conservation of some sites, such as the Point Pelee National Park and Long Point National Wildlife Area on Lake Erie, has been achieved through either federal, provincial or state action, but many other ecologically important sites remain at risk.

Joint initiatives by both non-government and government organizations have resulted in the Great Lakes Wetlands Environmental Agenda, which is directed at wetlands in both Canada and the United States. Also, agencies in Canada are currently developing a Great Lakes Action Plan which will focus on securing, restoring and conserving the Canadian wetland sites that are at greatest risk.

The Great Plain of the Koukdjuak

The Dewey Soper Migratory Bird Sanctuary is located on the Great Plain of the Koukdjuak on the west coast of Baffin Island, North West Territories and covers 8,160 square kilometres (3,150 square miles). The plain is an extensive arctic oasis for wildlife, with sedge lowland and peaty deposits found throughout the region. It has a unique karst landscape of circular shallow ponds, raised beaches and shallow streams overlying a limestone bedrock. A low relief shoreline results in ice-scoured tidal flats stretching 15 kilometres (9 miles) into the sea.

Designated a Ramsar Site in 1982, the Plain supports the world's largest goose colony. In July and August, as many as 230,000 pairs of lesser snow geese (*Anser coerulescens*) may nest here. Canada geese (*Branta canadensis*), oldsquaw duck (*Clangula hyemalis*), king eider (*Somateria spectabilis*) – the bill of which is considered an aphrodisiac in Greenland – and many shorebirds also nest in or use the area. A major caribou herd migrates to and from the plain each season.

Mary's Point and Shepody Bay

The extensive mud flats and tidal marshes at the head of the Bay of Fundy, in New Brunswick, eastern Canada, form one of the most important resting and feeding areas for shorebirds in North America. The four Ramsar Sites located here include the Mary's Point and Shepody Bay areas, designated under the Convention in 1982 and 1987 respectively. In addition, by 1987, both sites were incorporated in the Western Hemisphere Shorebird Reserve Network (WHSRN).

The sites attract a wide variety of shorebirds, particularly the semi-palmated sandpiper (*Calidris pusilla*), of which more than 200,000 may be present on a single day in late summer; over a million sandpipers pass through Mary's Point between July and September. Other species using these sites include least sandpipers (*Calidris minutilla*), short-billed dowitchers (*Limnodromus griseus*), black-billed plovers (*Pluvialis squatarola*) and red knots (*Calidris canutus*). The chief food supply on the mud flats is mud shrimp (*Corophium volutator*), which,

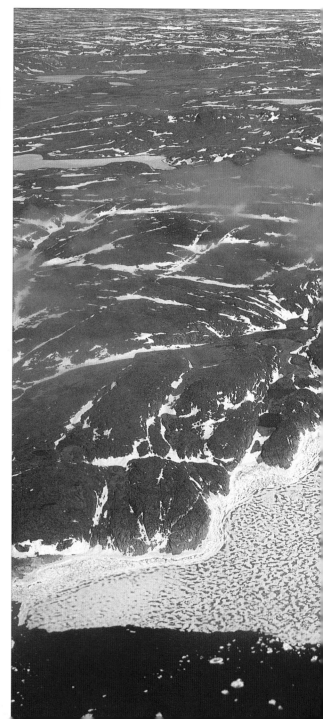

although common throughout European coastal areas, are found only in this region of North America. These habitats are irreplaceable without them sandpiper populations would not be able to store the invaluable fat and protein essential for their lengthy migration to overwintering havens in Surinam, northwest South America, and other parts of South America. Similar coastal sites lying in the Bay of Fundy are used by a wide range of waterfowl during the spring migration.

Left **A broadwalk allows visitors to stroll across the marshes at Point Pelee National Park in Ontario, Canada. Point Pelee National Park is one of the success stories of the Canadian conservation movement. In other parts of the country, however, stretches of similar coastal wetland are under threat from housing and recreational developments.**

James Bay Hydro Development

Proposals by three provincial power authorities exist for many of the major rivers flowing into the Hudson-James Bay region. The largest is the La Grande River Project of Hydro Quebec, which proposes the construction of additional major hydro reservoirs and flooding thousands of square kilometres of upland and river habitat. The major facilities already present have had adverse effects on life in the southern region of James Bay and the eastern coast of Hudson Bay. Large areas flooded in the 1970s and 1980s, for hydroelectrical storage in the James Bay Development Region of Quebec resulted in the release of large quantities into the reservoir waters of mercury from natural estuarine deposits. The mercury had major impacts on wildlife and human populations downstream.

The current proposals will have serious consequences for the extensive wetlands found alongside many of the region's rivers and estuaries and are strongly opposed by many environmental groups. They object to the project on the grounds that it will reduce the habitat available for fishery and other wildlife resources and alter the prosperous hunting and trapping areas for local people, so bringing substantial changes to their livelihoods and cultures.

Left **Low-lying peatland stretches inland from the coast at Cape Dorset on Baffin Bay. Freezing water below ground creates patterns of ridges. In the High Arctic, where polar desert predominates, wetlands provide oases of water and vegetation.**

Above **Arctic fox (*Alopex lagopus*) at Cape Churchill, Hudson Bay. The Hudson-James Bay coastal wetland is the world's longest marshy coastline. Although in winter it lies under snow and ice, the spring thaw turns the region into a rich wetland habitat.**

Western Canada and Alaska

The landscapes of Alaska and the neighbouring provinces and territories of western Canada contain most of North America's remaining wilderness areas. A rich diversity of subarctic, boreal, mountain, prairie and oceanic wetlands make up much of this wilderness. Taken together, western Canada and Alaska contain approximately 1.4 million square kilometres (550,000 square miles) of wetlands, an estimated 25 per cent of the world's total.

Large estuarine deltas and their associated salt marshes and tidal flats dominate the coastal wetlands. Some of the most extensive salt marshes lie along the 800-kilometre (500-mile) shoreline of the Alaskan Yukon-Kuskokwim Delta, itself one of the largest deltas in the world. To the seaward side of the marshes lie sand and mud flats, which are in places over 10 kilometres (6 miles) wide; in total they cover about 530 square kilometres (200 square miles).

1 Yukon-Kuskokwim Delta

The coastal plain of the Yukon-Kuskokwim Delta is one of the most productive wildlife areas in Alaska. It supports some very high densities of waterfowl and shorebirds. In some areas, during late summer and autumn, as many as 20 species of shorebird use the intertidal zone, often found in densities exceeding 15 birds per hectare (37 per acre). Coastal estuaries and rivers are particularly important to the large flocks of moulting birds and broods of brant, emperor, cackling and white-fronted geese (*Anser albifrons*).

The Copper River Delta, which lies just east of Prince William Sound, supports a variety of wetlands, including sandy barrier islands, estuarine tidal flats, salt marshes and freshwater marshes. The delta is an important staging area in the migration of numerous species of waterfowl and shorebirds. It is also a major nesting area for the dusky Canada goose (*Branta canadensis*) and for trumpeter swans (*Cygnus buccinator*).

Cold haven

Algae and vascular aquatic plants such as eelgrass (*Zostera marina*) commonly dominate the intertidal flats. Eelgrass favours the soft sediments of shallow, protected lagoons and will not grow in large river deltas, glacial fjords or most arctic environments. The Izembek lagoon on the Alaska Peninsula contains one of the largest eel grass beds in the world, at over 400 square kilometres (150 square miles). In 1986, this area was included by the United States in the Ramsar Convention's list of wetlands of international importance. Most of the world's emperor geese (*Philacte canagica*) and nearly all of the brent goose, or brant (*Branta bernicla*), that fly over the Pacific use the lagoon during the spring and autumn migrations.

Along the Arctic coast of Canada, and the Arctic and Bering coasts of Alaska, wetlands are currently under little developmental pressure and remain largely untouched; but economic decisions may open up the Arctic National Wildlife Refuge, with far-reaching consequences for the landscape. Along the Pacific coast, however, human populations are concentrated next to wetlands along the narrow coastal strip. In Juneau, Alaska, 13 per cent of the wetlands have been filled for commercial and residential development since 1948.

Further south in Canada's Fraser River Delta, estuarine wetlands are also under intense development pressure. In winter, this delta supports the highest densities of waterfowl, shorebirds and birds of prey anywhere in Canada. In the summer, up to a million shorebirds of 24 species feed and rest in the delta. However, these wetlands compete with the needs of Vancouver. From 1976 to 1982, over 28 per cent of the wetlands in the Fraser lowland were developed for a variety of other uses.

Freshwater wetlands

Inland, freshwater boreal wetlands cover an estimated 600,000 square kilometres (230,000 square miles) of Alaska, and over 20 per cent of central Canada. They include a wide range of wetland types, including peat bogs, fens, delta marshes, floodplain swamps and both wet tundra (which has areas of standing water) and moist tundra (with no areas of standing water). Peat is the dominant soil type in these wetlands, which contain a major portion of the world's carbon in the form of the extensive peat accumulations.

Moist and wet tundra underlain by permafrost (permanently frozen soil) cover vast areas in northern Canada and Alaska. These treeless landscapes are dominated by wetland grasses, sedges and dwarf shrubs. The common mare's tail (*Hippuris vulgaris*) and pendent grass (*Arctophila fulva*) are common in the wet tundra areas. Tussock cotton grass (*Eriophorum vaginatum*) and low shrubs such as dwarf birch (*Betula nana*) and bog blueberry (*Vaccinium uliginosum*) dominate the moist tundra. South of the tundra region and away from the coast, black spruce (*Picea mariana*) muskeg covers hundreds of thousands of square kilometres. Tamarack (*Larix laricina*) is associated with the black spruce in wet lowland sites. Common plants in the marshes of the region include water sedge (*Carex aquatilis*), marsh cinqefoil (*Potentilla palustris*) and bluejoint grass (*Calamagrostis canadensis*).

In the Pacific coast forest region, bogs and muskegs predominate where conditions are too wet for tree growth. The vegetation in these wetlands consists of *Sphagnum* mosses, sedges, rushes, shrubs and stunted lodgepole (*Pinus contorta latifolia*) pine. Western hemlock (*Tsuga heterophylla*) or Alaska cedar (*Chamaccyparis nootkatensis*) dominate the forested wetland sites.

The majority of the freshwater wetlands of the region are subjected to little pressure from development. In Alaska, approximately 800 square kilometres (310 square miles) of freshwater wetlands have been lost since colonial times – about 0.1 per cent of the original area. Until now, very minor areas of Canadian boreal wetlands have been drained to improve forest productivity or developed for other purposes.

Queen Elizabeth Islands

Polar Bear Pass

Victoria Island

Rasmussen Lowlands

3 Queen Maud Gulf Migratory Bird Sanctuary

NORTHWEST TERRITORIES

McConnell River

Whooping Crane Summer Range

Lake Athabasca

Fort Chipewyan
Peace-Athabasca Delta

MANITOBA

Lake Winnipeg

ASKATCHEWAN
Quill Lakes

Last Mountain Lake Winnipeg
Delta Marsh **Oak-Hammock Marsh**

U S A

Ramsar Sites
Parks and Reserves
Water Bodies
25-50% Wetland
50-100% Wetland
Prairie Pothole Region
Important Bird Areas

2 Copper River Delta

In 1964, the entire Copper River Delta was lifted 2–4m (6.5–13 ft) as a result of a massive earthquake which reached 8.5 on the Richter scale. Tidally influenced, sedge-dominated marsh has extended seaward by as much as 1.5 km (1 mile) across tidal flats, while pre-earthquake salt marshes have converted to freshwater systems dominated by shrubs and emergent plants.

3 Ramsar Sites

Fourteen Ramsar Sites are located in western Canada, including the largest Ramsar Site in the world, the Queen Maud Gulf Migratory Bird Sanctuary. This vast sanctuary covers about 62,800 sq km (24,000 sq miles), an area almost the size of Ireland.

The North American Waterfowl Management Plan

The severity of wetland loss and the key role wetlands play in the western agricultural economy has become widely recognized. As a result, Canada, the United States and Mexico have joined forces to create an innovative conservation programme, the North American Waterfowl Management Plan, the aim of which is to conserve the continent's waterbirds. The plan is administered by the Canadian Wildlife Service and the United States Fish and Wildlife Service. As part of this plan, Canadian and American federal and provincial government, and private-sector partners, implemented several joint venture programmes designed to conserve prairie wetlands in the hope that this will encourage the re-establishment of waterfowl and other wildlife populations, as well as promote soil and water conservation.

Local programmes

The aim of the long-term programme is to manage 240,000 square kilometres (93,000 square miles) across Canada, of which 80 per cent are in western Canada, and to protect and conserve 4,500 square kilometres (1,700 square miles) of waterfowl habitat in the Prairie Pothole region of the United States. In the first few years, from mid-1988 to early 1991, over 550 square kilometres (200 square miles) of wetland and upland habitat had been secured in western Canada through private landowner stewardship and acquisition projects. There are approximately 200 other regional and local wetland projects associated with the programme, such as the Heritage Marsh Programme (see below) and Prairie Care programmes that are run by provincial organizations as well as non-governmental organizations such as Ducks Unlimited Canada. As well as the actual wetland management aspect of the programme, a great deal of attention is being given to producing and dispensing education material, and carrying out research of individual species of special concern.

The Province of Alberta is currently developing, through a public consultation process, a provincial wetland conservation and management policy. This process is being mirrored by a similar policy consultation process in Saskatchewan. Manitoba and Saskatchewan are also addressing questions of soil, water and agricultural conservation programmes.

Below left **Prince William Sound, Alaska,** the northern end of a belt of intermittent coastal wetland that extends south to Seattle. Among the many non-aquatic birds that breed in the region are bald eagles (*Haliaeetus leucocephalus*), ospreys (*Pandion haliaetus*), and short-eared owls (*Asio flammeus*).

Below **Barrier islands in the delta of the Copper River.** Tidally influenced wetlands such are these are particularly vulnerable to sea-borne pollution. Had the *Exxon Valdez* oil-spill, for example, occurred further down the coast, important delta and estuarine wetland habitats would have been devastated.

Left **An emperor goose** (*Anser canagicus*) protects its young by using its wings in a dramatic threatening display. A relative of the snow goose, the emperor goose breeds in western Alaska and northeastern Siberia, feeding among seaweed beds and estuarine mud flats.

Heritage Marshes, Manitoba

The Heritage Marsh Programme is an innovative conservation programme designed to bring together the strengths of the Government of Manitoba with those of non-governmental organizations such as Ducks Unlimited Canada and the Manitoba Wildlife Federation. The programme targets prairie marshes which serve as critical waterfowl moulting and staging areas, and provide refuge for other waterbirds during western Canada's drought periods.

Despite their large size and permanent nature, the Heritage Marshes are threatened by agricultural encroachment and flooding by hydroelectric projects. To address, and hopefully fend off, these threats the programme works with local municipalities and landowners to protect and enhance the values of critical sites.

Peace-Athabasca Delta

Stretching over 3,200 square kilometres (1,200 square miles), the Peace-Athabasca Delta in Alberta provides vistas of vast sedge meadows and shallow lakes. Their lush, emergent vegetation provides an abundant food supply for wildlife, including the free-ranging bison (*Bison bison*) and a rich array of other mammal species such as grey wolves (*Canis lupus*) and lynx (*Felis lynx*). In spring and autumn over a million birds, including many species of ducks and geese and the endangered whooping crane (*Grus americana*), use the delta.

Altered environment

Despite inclusion within Canada's biggest national park – Wood Buffalo – and a listing under the Ramsar and World Heritage Conventions, the Peace-Athabasca Delta is a striking example of the frustrating effects of external factors for even a large-scale wetland conservation initiative. In 1969, completion of the Bennet Dam for hydroelectrical storage vastly reduced water flows to the delta. Park officials estimate that, to date, as much as 25 per cent of the delta has been affected and this will rise to 80 per cent if action is not taken in the next 40 years. These changes have been of special concern for the Fort Chipewyan Indians, whose lives centre on the delta. In the words of one Indian, "Now the lakes where we used to trap are all poplar and willow. Over 100,000 to 300,000 muskrats came out of the delta every season. I used to get 50, 60, up to 100 muskrats a day. Only one year did I have to go on social assistance." Today, the number of muskrats trapped has dropped to 50,000 to 60,000 a year.

On top of these impacts, the delta is threatened by water pollution from paper mills currently under construction in British Columbia and proposed in Alberta. And in 1990, proposals were made for the extermination of the bison population due to the risk of tuberculosis and brucellosis spreading to other populations and cattle outside the park.

Prairie Potholes

The prairies which stretch across the southern third of Alberta, Saskatchewan and Manitoba are a semi-arid, generally treeless environment dotted with over 4 million small wetlands, known as potholes. This mosaic of wetlands continues to the south in the American states of Iowa, Minnesota, North Dakota, South Dakota and Montana. Referred to as the "duck factory" of North America, this area is critically important to breeding waterfowl. Of a total of 777,000 square kilometres (300,000 square miles), about 65 per cent of the Prairie Pothole region lies in Canada; 274,000 square kilometres (106,000 square miles) are in the United States.

As the glaciers receded from this part of the continent 10,000 years ago, they left behind them small depressions in the landscape created by the scouring action of the ice. These "potholes" now support a variety of small wetlands, ranging from wet meadows and shallow water ponds, to saline lakes, marshes and fens. The vast majority of these wetlands collect water in the form of rain and melting snow, very few have surface inlets or outlets. The density of potholes may be as high as 60 sites every square kilometre (155 per square mile).

As rainfall is seasonal, many wetlands only have water for the spring and early summer periods. At this time they are flush with colour as flowers bloom and birds breed. These wetlands play a vital role in the maintenance of nearly all forms of prairie wildlife, while also performing a wide range of other functions. Depending upon their location, basin shape and size, pothole wetlands can store floodwaters, absorb nutrients, recharge groundwater, and provide water and forage for domestic animals.

Agricultural pressure

The prairie landscape is, however, under intensive agricultural use for grain production. Cropland makes up about 68 per cent of the prairie pothole region of the United States and Canada. Several technological events in modern history have spurred the agricultural development of the prairie lands. As early as 1930, the advent of the mechanized tractor eliminated the need to produce feed for mules and horses. Much of the rangeland that had been devoted to pasture was converted to cash crops with the result that between 1930 and the mid-1960s the total area of pasture land dropped from 263,000 square kilometres (100,000 square miles) to 28,000 square kilometres (11,000 square miles). Furthermore, since the 1960s, farm size, equipment size and the use of irrigation have all increased. These, and other intensive farming practices have encouraged further drainage of wetland habitats and the use of uplands immediately adjacent to wetlands.

Vanishing potholes

The extent of prairie pothole wetlands has declined dramatically since settlement of the region in the 1850s. At that time, wetlands covered 16–18 per cent of the prairie regions of the States of Minnesota, South Dakota, North Dakota and Iowa. Today, wetlands cover only 8 per cent of this region. In Iowa, the area of wetlands was reduced from more than 16,000 square kilometres (6,000 square miles) prior to 1875 to 3,800 square kilometres (1,450 square

Below **A wintry, aerial view showing the prairie pothole terrain south of Minnedosa in southern Manitoba, Canada. The "potholes", which vary dramatically in size, depth, shape and density, are the physical remains of the ice sheets that once covered this region of Canada and the United States. As the ice retreated, it scoured the land, creating thousands upon thousands of shallow depressions. The roads that form a cross in the top half of the picture give some indication of the scale of the region and the number of the potholes.**

Right **Just one example of the many endangered potholes that are fast being drained in Alberta in Canada. Modern farming equipment and methods have been responsible for the destruction on a massive scale of these miniature wetland environments.**

Bottom right **Nicknamed the "duck factory", the Prairie Pothole region provides breeding areas for many species of waterfowl. The birds build their nests in the fringing marshes and feed on the potholes' mass of aquatic vegetation.**

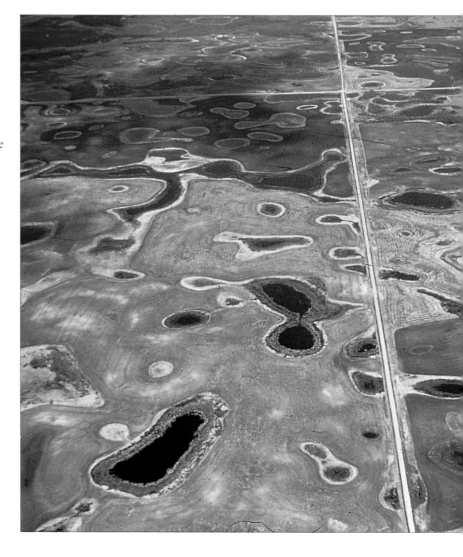

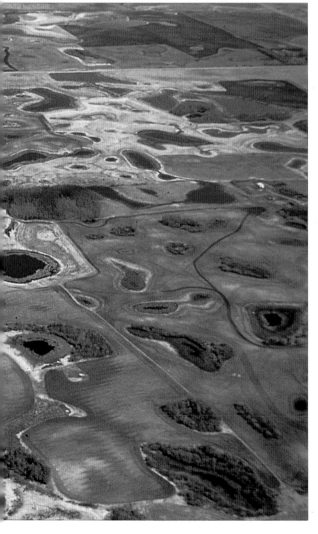

miles) in 1906, and 1,500 square kilometres (580 square miles) by 1922. Today in Iowa, those areas of wetland that still remain are only relicts and most are protected under some form of state or federal stewardship.

In Canada, the patterns of drainage has been repeated since settlement of the prairies in the early decades of this century. Many more small wetlands have been degraded by land use practices, such as water pollution from fertilizer and pesticide use, soil erosion and road construction. Together these practices have resulted in losses of as much as 71 per cent of the original wetland area over the last 70 years.

The United States – The Lower "Forty-Eight"

When the British colonists established the first permanent settlement on the shores of Chesapeake Bay in 1607, the area that now makes up the lower 48 states comprised over 890,000 square kilometres (340,000 square miles) of wetlands. Today, the United States Fish and Wildlife Service (USFWS) estimates that only just over 415,000 square kilometres (160,000 square miles) remain. The processes of agricultural, industrial and urban development which dominated the 385 years since 1607 have been the major causes of American wetland loss.

Today, however, wetlands have become recognized as an invaluable public resource. Roughly two-thirds of the commercially important fish and shellfish species harvested along the Atlantic and Gulf coasts, and half of the Pacific coast species, are dependent on estuarine wetland habitats for food, spawning and/or nursery areas. And 60 to 70 per cent of the 10 to 12 million waterfowl which nest in the lower 48 states do so in the Prairie Pothole region of the upper mid-west. Millions of additional shorebirds, egrets, herons, terns, gulls, pelicans and other waterbirds depend upon a range of wetlands throughout the country.

The east coast of the United States presents a landscape of fresh- and saltwater marshes running into the estuaries dotted along the frequently highly indented coast. In the southeast, tidal wetlands are especially abundant in large, drowned river valley estuaries such as Delaware and Chesapeake bays. These estuaries are major resting and wintering grounds for the thousands of waterbirds that breed on the wetlands further north. They also provide a habitat for soft-shelled crabs, oysters and shrimp, and many fin fish, all of which yield important harvests and form an important part of the economy of the southeast.

Inland from the coast, the northeast is a land of freshwater bogs, fens and some extensive peatlands, with forested swamps as the single most important freshwater wetland.

This pattern is mirrored in the unglaciated south. Here, extensive deciduous forested swamps, known as the "bottomland hardwoods", occupy the floodplains of river courses and large coastal basins which are flooded annually by slow-moving streams. The Great Dismal Swamp of Virginia and North Carolina, and the Okefenokee Swamp in Georgia are among the largest and best known. Freshwater, bog-like wetlands called *pocosins* and Carolina bays are found on the coastal plain of the mid-Atlantic States.

Coast to coast

On the Florida Peninsula, fresh water originating from the north of Lake Okeechobee in central Florida flows south in wide, flat, shallow basins through the Great Cypress Swamp and the Everglades, the largest marsh system in the United States. The flow continues into the Gulf of Mexico, where fringing mangroves are North America's only forested marine wetlands. Around the Gulf Coast of Mississippi and Louisiana, to the west, lie huge freshwater marshes and forested wetlands that have developed in the ancient

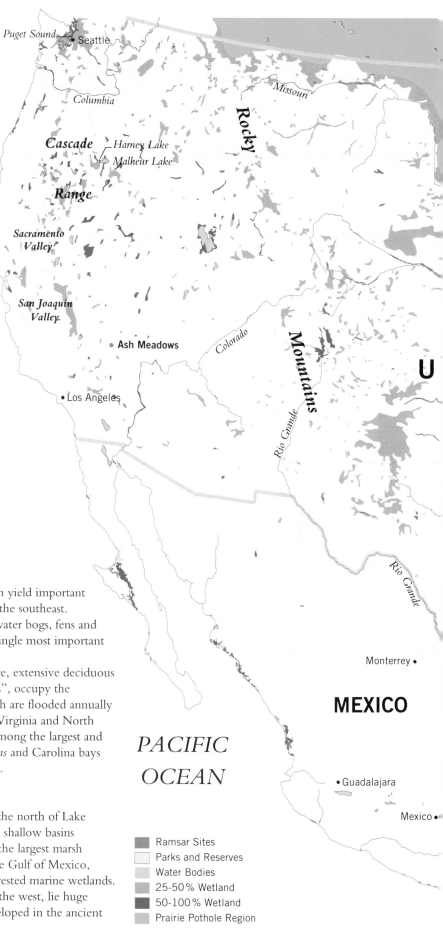

Ramsar Sites	
Parks and Reserves	
Water Bodies	
25-50% Wetland	
50-100% Wetland	
Prairie Pothole Region	

deltas of the Mississippi River. These wetlands provide migratory and wintering habitats for many waterbirds, including several million ducks. In Texas, saline lagoons dominate much of the state's coastline.

Coastal marshes are less extensive along the Pacific coast and are located primarily in estuaries in the states of Oregon and Washington. Inland, the freshwater wetlands of the river basins and agricultural areas are important wintering areas for dabbling and diving ducks, geese and swans. However, these wetlands cover just a fraction of their former extent, and although California is the major waterfowl wintering area of the Pacific coast, the vast freshwater wetlands that once sat at the confluence of the Sacramento and San Joaquin rivers have been drained for agriculture. To the north in the Cascade mountain range, wetlands occur on narrow floodplains and on isolated alpine and subalpine meadows. Along the coast, tidal marshes fringe the rivers that flow into the Pacific Ocean. These wetlands once extended well inland, but are now diked for pastures. The estuary of the Columbia River and Puget Sound, however, still contain important tidal wetlands.

Lowland floodplains to glacial lakes

In the vast continent that lies between America's coasts, the wetland landscape changes from one of bogs, fens and forested swamps in the Great Lakes Basin to the Prairie Potholes of the upper Mississippi. Further south, wetlands occur primarily on the lowland floodplains and shifting river channels. Along the lower Mississippi, wetlands are inundated seasonally, while permanent wetlands occur in small floodplain ponds. These wetlands constitute the prime wintering area for waterfowl in the central part of the United States, and provide stop-over and resting areas in the single most important waterfowl migration route in North America.

To the west, wetlands are less abundant. In the Rocky Mountains, alpine wetlands are found on glacial lakes, and along slow-flowing streams in glaciated valleys. On the Columbia River plateau, wetlands are restricted to the river course, internal drainages, potholes and steep-sided lakes. Extensive marshes bordering the alkaline Harney and brackish Malheur lakes make up one of the most extensive inland marsh systems of the lower "forty-eight".

"No Net Loss"

In 1987, as a response to the dramatic and continued loss of wetlands, the National Wetland Policy Forum established by the Environment Protection Agency and the Conservation Foundation recommended that the nation adopt the immediate goal of "no net loss" of wetlands and a long-term goal of actually increasing the nation's wetlands.

Although the no net loss goal recognizes that there will continue to be some losses due to the need for projects that are critical to the public interest, the National Wetland Policy Forum argues that those losses should be balanced by restoring former wetlands or creating new ones.

Into practice

No net loss has been well received as an objective. It is being adopted as an operational guideline by several federal agencies and states, as well as a number of municipalities who have their own wetland regulatory programmes. But implementing the goal has become increasingly complicated because of an effort to change the federal procedure by which the boundaries of wetlands are determined. A procedure that has the backing of oil, agricultural and home-builder interests, but denounced by the science community, was proposed by government officials. Its adoption, according to some scientists, would result in approximately half of the remaining wetland areas in the United States being removed from federal wetland regulatory protection.

California's Central Valley

In the mid–1800s, the rivers of the California Central Valley flooded each winter to create vast seasonal wetlands in the valley floor and delta. In the 1850s, these wetlands amounted to more than 160,000 square kilometres (62,000 square miles). Vast flocks of migratory and resident waterbirds used these wetlands and rivers, which also provided spawning and rearing habitats for salmon and other species of fish. However, the 1850s saw the first widespread conversion of the wetlands when farmers diked the floodplains of the Sacramento Valley for cultivation.

Destructive activities

By 1939, 85 per cent of the original wetland area had been lost, modified largely by levee and drainage activities, and local water diversion projects. Significant wetland losses have continued, with a further 20,000 square kilometres (8,000 square miles) of vegetated wetlands converted between 1939 and 1985. Agriculture was responsible for about 95 per cent of the net loss of these wetlands. Today, there are an estimated 1,500 square kilometres (580 square miles) of all wetland types remaining in the Central Valley of California. The majority of these are managed areas that have been created or maintained by controlling water.

The wetlands of California's Central Valley, as other wetlands of the American West, not only compete for space with agriculture but also for water. There is a continuing thrust to develop the finite supply of water for irrigated agriculture and urban–industrial uses without an adequate, guaranteed, clean water supply for public and private wetland areas.

Above **Sunset over the tidal marshes of Chesapeake Bay, Maryland, USA – the largest estuary in the country. This immense system of fresh and saltwater marshes is highly productive, and some 90 per cent of the nation's annual catch of striped bass (*Roccus saxatilis*) is taken from local waters, together with large numbers of crabs and shellfish. However, the long-term productivity of these wetlands is threatened by industrial and domestic pollution.**

Right **Map showing a section of the Prairie Pothole region near Tuttle in North Dakota, USA. Each pothole has been meticulously coded. Prepared as part of the National Wetlands Inventory programme, such maps provide detailed information that is invaluable in assessing the impact of proposed drainage, land-fill schemes or other forms of wetland conversion, and in developing an integrated wetlands conservation strategy for the future.**

Chesapeake Bay

On the northwest Atlantic coast of the United States, the Susquehanna River ends at Chesapeake Bay. The largest estuary in the United States, Chesapeake Bay holds more than 82,000 billion litres (18,000 billion gallons) of water and is one of the most productive ecosystems in the world. It has an average annual production of 13 million kilograms (28 million pounds) of oyster meat and 25 million kilograms (55 million pounds) of blue crab meat.
In 1980, the economic value of seafood, sport-fishing and related activities centred around Chesapeake Bay was valued at about US$756 million. In addition, the area enjoys immense biodiversity, with more than 2,700 plant and animal species living in the bay's waters and wetlands.

Chesapeake Bay receives its fresh water from a 166,000-square kilometre (64,000-square mile) watershed that comprises six states. It extends from the head of the bay near Baltimore, in Maryland, through Pennsylvania to the south of New York State. By 1992, more than 13 million people were living within this area, which contains over 150 rivers and streams.

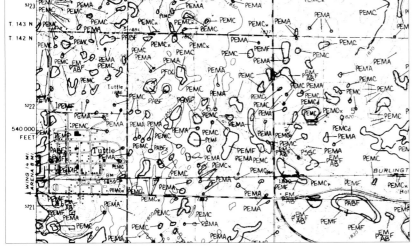

The municipal wastes from this watershed and the commercial ships and pleasure boats that use the bay have had a major impact on this vast ecosystem. Similarly, land use within the watershed has been shifting from forest and agriculture to urban and residential uses. Between 1950 and 1980, the amount of developed land rose from 5 per cent to 15 per cent.

The effects of increased urbanization and farming have resulted in serious deterioration of water quality and productivity of fish and other aquatic organisms. Widespread losses of freshwater wetlands in the watershed area and of estuarine wetlands in the bay itself have become a matter of regional concern.

National Wetlands Inventory

In response to a growing concern for the loss of America's wetlands, the United States Fish and Wildlife Service (USFWS) in 1975 began the United States National Wetlands Inventory. By creating detailed wetland maps for each state, and documenting and analysing historical status and trends in wetland resources, the project is designed to provide an information base upon which future United States wetland conservation policy can be based.

Using a combination of aerial photographs, soil data, topographic maps and other regional studies to identify and classify wetlands half a hectare (an acre) in size or larger, the National Wetlands Inventory produces large-scale maps that show the location, shape and characteristics of wetlands and deepwater habitats. These are then overlaid on a topographic base map.

The first maps were produced in 1980, and by the end of 1991 the Inventory had completed mapping for 75 per cent of the lower 48 states. Hawaii, Guam, Puerto Rico and about 25 per cent of Alaska's wetlands have also been mapped.

Left **Geometrically precise farmland in the southern part of California's Central Valley. Less than 150 years ago, the valley floor contained vast tracts of seasonal marshland;** today less than 1 per cent of the original wetland remains, the rest has been drained to grow crops which depend on irrigation water "imported" from neighbouring states.

Mississippi Delta and the Bottomland Hardwoods

The southern states of the United States support some of the country's most important wetland resources. The Mississippi Delta alone contains about 40 per cent of all coastal wetlands in the lower 48 states. Equally important, but more widespread in their distribution, are the bottomland hardwood wetlands which characterize the floodplains flanking the major streams of the southeast of the country.

As in many other regions of the world, these wetlands have seen many changes over thousands of years, but in recent decades the changes have become more and more dramatic. For example, over the past 7,000 years, a total of 36,000 square kilometres (14,000 square miles) of land have been built up in the Mississippi Delta, and because the path of the river has changed as a result of natural processes some 18,000 square kilometres (7,000 square miles) of this land has been destroyed. Today, the situation is very different, with an annual loss of 91 square kilometres (35 square miles) of the delta being due primarily to the leveeing of the river. In the past, the sediment carried by the river was deposited in the delta, but the modified river now carries this far offshore into the deep water of the Gulf of Mexico.

In the lower Mississippi valley, more than 2,500 kilometres (1,500 miles) of dikes have been built and the river is confined for more than 1,000 kilometres (600 miles) of its course. In addition to shortening the river, increasing stream velocities and contributing to the erosion of the delta, this engineering work now prevents natural flooding of the floodplain forested wetlands. These once covered 100,000 square kilometres (39,000 square miles), but now only 18,000 square kilometres (7,000 square miles) remain; the rest was cleared to provide land for agriculture.

It was the desire for economic development that brought about these large-scale developments. However, the economic costs, along with the investments required to maintain the Mississippi Delta, and the loss of the natural flood-retention qualities of the bottomland hardwoods, is only now being widely realized. It may already be too late for many of the larger mammals, such as jaguars (*Panthera onca*) and black bears (*Ursus americanus*), whose populations in these last refuges have been severely diminished.

"Reclaiming" land for agriculture

While the Mississippi symbolizes the impact of a century of intensive development upon the nation's wetland resources, many other wetland systems are equally threatened. Only 35 per cent or less of the bottomland hardwood forest which existed when white settlers entered the area remains. Again, the major force behind this loss is agricultural development. Between the 1950s and 1970s, Louisiana, Arkansas and Mississippi lost vast areas of bottomland hardwoods to crop production, primarily soyabean and cotton. Agricultural subsidies encouraged wetland loss, and much of the money for draining and diking of the bottomland hardwoods has come from government subsidies. The inherent conflict in such government support for the widespread destruction of a unique resource is now being addressed. Systems of multiple use are being promoted, which make maximum sustainable use of the timber, fish and shellfish, wildlife and recreation values of the wetlands without affecting the valuable hydrological functions, such as water purification and flood retention.

Below **The muskrat (*Ondata zibethicus*) is one of the many animals that inhabit the wetlands around the mouth of the Mississippi River, and these traditional habitats are now threatened. Although not so important as previously, some 50 per cent of the United States fur harvest comes from this region, together with about a third of the nation's total seafood catch.**

Bottom **Sulphur refinery on reclaimed land in the Mississippi Delta, with a small boat marina in the background. Easy access by water through dredged channels has made the delta wetlands an attractive area for industrial and recreational development. The combination of pollution and habitat destruction has already made parts of the delta untenable for wildlife.**

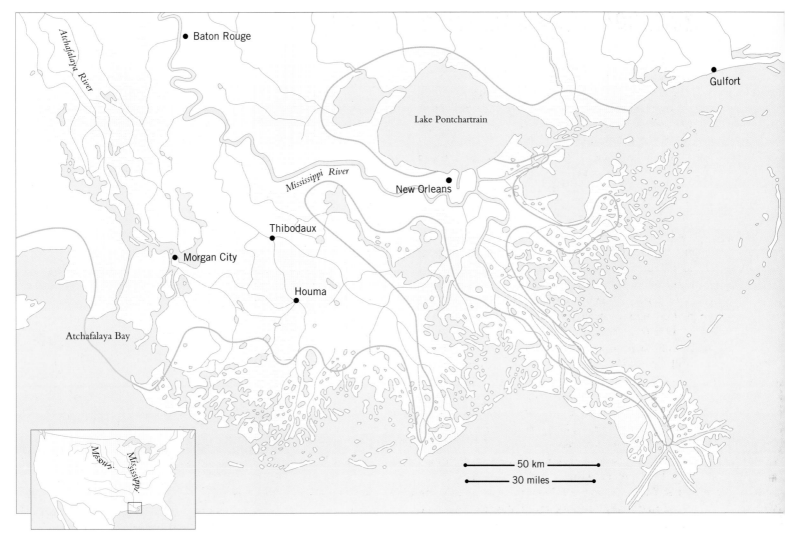

—— Wetland loss

—— Wetland gain

☐ Wetland area

Above **Flood defences along the banks of the lower Mississippi have increased water velocity so that sediment that was once deposited in the delta is now carried out to deep** water. As a result, the delta is shrinking rapidly, at a rate of about 100 sq km (40 sq miles) per year. The Gulf coastline will look very different in 50 years' time (shown by red line).

The productive hardwoods

The bottomland hardwood wetlands, characterized by bald cypress trees and still, duckweed-covered waters, once stretched over 370,000 square kilometres (140,000 square miles), but today cover less than 130,000 square kilometres (50,000 square miles). These breathtaking wetland systems support plant and animal communities that are highly adapted to fluctuating water levels and are closely coupled to the estuarine systems at the river mouths.

Such floodplain forests are highly productive. The annual floods supply the system with sediments, dissolved organic matter and nutrients. Over 50 species of fish feed or spawn on the floodplains, and their distribution and abundance in the rivers is dependent on the forests' flooding cycle. When flood control levees are built along the stream banks to prevent flooding and encourage conversion of the forest to other uses, this critical breeding habitat is lost.

At the mouths of many of the rivers that drain the bottomland hardwoods, there are estuaries that support highly productive and valuable marine fisheries. The same flooding that brings life-supporting elements to the floodplain forests also flushes organic material and silt from the forest floor and delivers it to the estuaries. This flushed material, some of which comes from the leaf and twig litter of the forests themselves, supports the estuaries' complex food chains. In this way, the annual flood maintains the economic viability of the valuable fish and shellfish industry of the river mouths.

The Everglades

The Florida Everglades comprises one of the largest freshwater marshes in the world. Once this wet wilderness covered 10,000 square kilometres (4,000 square miles), from Lake Okeechobee to the southwest tip of Florida. At that time the Everglades formed what the Indians who lived there called the "grassy waters", a river of grass creeping to the sea.

As the name suggests, the Everglades is dominated by sawgrass (*Cladium* sp.), which can form single stands covering hundreds of square metres and often reaching 4 metres (13 feet) in height. The Everglades also supports a great variety of aquatic plant communities, including lilies (*Nuphar* and *Nymphaea* spp.) and bladderworts (*Utricularia* sp.), which mix with the sawgrass in many areas. Where irregularities in the limestone bedrock have left raised areas, small islands or hummocks develop. These hummocks are populated by a diversity of hardwood trees and support varied populations of animals, including the *Liguus* tree snails, which are endemic to southern Florida and the islands of Cuba and Hispaniola.

North of the marsh lies Big Cypress Swamp, while to the south lie the fringing mangroves of Florida Bay. These forested wetlands complement the sawgrass and its associated communities. Thus, while many of the colonial waterbirds nest in the cypress and mangrove trees, they feed in the marsh. Alligators, on the other hand, confine themselves to the freshwater marshes, with the more rare American crocodiles living in salty and brackish waters.

The Miami Canal

For much of the nation's early history the Everglades was untouched by major development efforts. However, at the turn of the century Florida's governor initiated the first drainage project, which was completed in 1909. Known as the Miami Canal, it connected Lake Okeechobee to the Miami River and the sea. Severe flooding in 1926, 1928, 1947 and 1948 resulted in the Central and South Florida Control project, which involved the construction of almost 1,300 kilometres (800 miles) of levees and 800 kilometres (500 miles) of canals, further helping to drain the land.

Today, less than 50 per cent of the original Everglades remains, the rest has been drained and converted to agricultural or urban uses. Agriculture is a multi-billion dollar industry, occupying over 2,700 square kilometres (1,040 square miles) of drained, freshwater wetlands. The freely flowing

"river of grass" has been replaced by an intensively managed, multiple-purpose water control system containing over 3,000 kilometres (1,900 miles) of canals and levees, and 150 major water-control structures. Only a fifth of the original Everglades, the most downstream portion, is protected within the Everglades National Park. This remnant of the original system, however, faces a variety of threats from upstream. The watershed, which extends 300 kilometres (190 miles) to the north, is not protected, and water flowing into the Everglades may originate from, or pass through, agricultural or urban areas. As a result, the quantity, quality and distribution of water, all critical in maintaining an ecological balance, have been altered, which in turn has led to declines in the biological resources of the Everglades.

Hope for the future

The challenge that the remaining Everglades presents to scientists, engineers, planners and managers is whether or not it can coexist with, and be part of, an intricate public water-management system – one that has to accommodate the unchecked water supply and flood control needs of agriculture, economic growth, and a rapidly expanding regional human population of over 4 million. Thus, while much has been achieved recently, such as the Florida State Government's multi-million dollar "Save our Everglades" project and the injection of millions of dollars of federal funds, the battle continues today.

In 1989, the Federal Government sued the South Florida Water Management District over its failure to prevent degradation of wetland ecosystems in the Loxahatchee National Wildlife Refuge and the Everglades National Park. The deterioration was caused by polluted waters released from the agricultural lands which now occupy large areas of drained marsh. Several million dollars were spent on legal fees alone and the cost of the environmental restoration schemes proposed is likely to exceed US$500 million. In July 1991, the case against the Management District was settled out of court amid much publicity, but the legal wrangle continues unabated largely because of the financial implications for the farming interests upstream.

Far left The sea grass meadows off the Florida coast still attract marine green turtles (*Chelonia mydas*), although they are decreasing. A combination of pollution and changes in on-shore hydrology are reducing the size of the meadows. However, the greatest threat to the green turtle is still people. In many countries, both the meat and the eggs are considered delicacies.

Top "Rivers of Grass", a part of the vast expanse of sawgrass (*Cladium* sp.) that constitutes the "grassy waters" of the Everglades. In the distance can be seen one of the "hummocks" of higher ground that supports a stand of hardwood trees.

Left West of the Everglades lies Big Cypress Swamp, which comprises 4,000 sq km (1,500 sq miles) of primeval glory. Between the towering bald cypress (*Taxodium distichum*), a relative of the California redwood, are clusters of cypress "knees" which probably serve to keep the tree roots aerated. Although it is dominated by bald cypress, Big Cypress Swamp also supports other tree species, including pond dwarf cypress (*T. ascendens*), as well as slash pine, pond apple and willow on drier ground. Big Cypress is an important wildlife conservation area, and it is thought to contain some of the country's last remaining black bears.

Mexico, Central America and the Caribbean

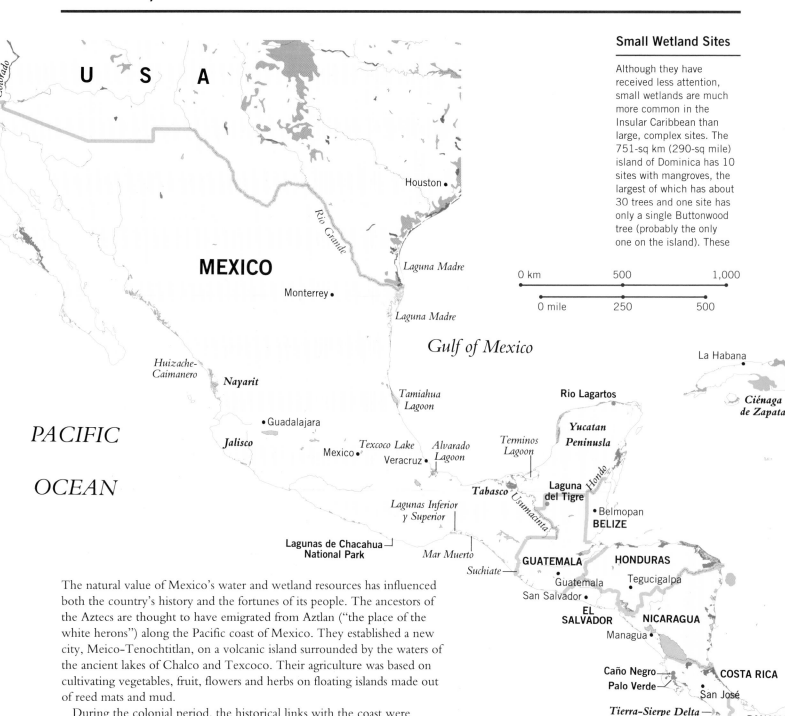

0 km		500		1,000
0 mile	250		500	

The natural value of Mexico's water and wetland resources has influenced both the country's history and the fortunes of its people. The ancestors of the Aztecs are thought to have emigrated from Aztlan ("the place of the white herons") along the Pacific coast of Mexico. They established a new city, Meico–Tenochtitlan, on a volcanic island surrounded by the waters of the ancient lakes of Chalco and Texcoco. Their agriculture was based on cultivating vegetables, fruit, flowers and herbs on floating islands made out of reed mats and mud.

During the colonial period, the historical links with the coast were weakened, and the European immigrants concentrated upon inland agriculture, mining and lumber harvest. By heading inland, the immigrants avoided the hot, humid disease- and insect-plagued coastal areas. Although this development placed intense pressure on inland wetlands – one of the "conquistadores'" main concerns was to drain the lakes surrounding Tenochtitlan – the coastal areas remained largely untouched. The 1980s, however, have seen increased coastal developments, mainly in fisheries, offshore oil exploitation and tourism.

Wetland wealth

Today, there are some 6,500 square kilometres (2,500 square miles) of inland wetlands, mainly lakes, lagoons and rivers. The country's main wetlands, however, lie on the coast, covering an area of some 12,500 square kilometres (4,800 square miles). Many of Mexico's wetland species are rare

- ■ Ramsar Sites
- ▨ Parks and Reserves
- ■ Water Bodies
- ■ Mangroves
- ■ Estuaries, Coastal Wetlands
- ■ Pans, Seasonal Wetlands
- ■ Swamp Forest, Flooded Forest
- ■ Predominantly Wetlands
- ■ Lower Density of Wetlands
- ■ Soda Lakes
- ■ Freshwater Marsh, Floodplains

Machalilla

sites have very little value in terms of exploitative resources, but their contribution to island biodiversity makes them worthy of preservation. There are only three wetland sites in the 98-sq km (40-sq mile) island of Montserrat, one of which has been declared a bird sanctuary. This site in Fox's Bay covers only 6 hectares (15 acres), with a restricted flora and fauna, and is surrounded by agricultural and residential land. Although small wetlands may seem less worthy of protection than the larger, more diverse and species-rich areas, the preservation of a network of small wetland sites throughout the West Indian island archipelago is of great significance to migratory bird flyways, as well as providing the only available habitat for those species of resident waterfowl that live here.

or endangered, notably freshwater and marine turtles, manatee, caiman, crocodiles, several species of fish and many resident and migratory birds. Waterbirds are especially well studied, and critical sites well documented. The Pacific coast is especially important for wintering brown pelican (*Pelecanus occidentalis*), brent goose (*Branta bernicla*), great blue heron (*Ardea herodias*), and many species of shorebird.

The major threats to Mexico's wetlands are agricultural and industrial development. Yet coastal lagoons and estuaries, which cover 16,000 square kilometres (6,000 square miles), are capable of producing food equivalent to that produced on 160,000 square kilometres (62,000 square miles) of Mexico's agricultural land. Lagoons such as Alvarado, Terminos and the wetlands in Tabasco in the Gulf of Mexico, and Huizache–Caimanero and the coastal areas of Nayarit in the Pacific, have the highest productivity of any habitat type in Mexico. If adequately managed they could in a year produce close to 180 kilograms of oysters per hectare (160 pounds per acre), over 10 times the amount of beef produced on drained wetlands. Yet wetland loss continues at an alarming rate.

The Caribbean

The Insular Caribbean is generally considered to consist of some 23 nation states, spread over 3,000 individual islands, cays and islets. The majority border the Caribbean Sea, although the Bahamas, Turks and Caicos Islands, and Bermuda to the northeast are included with the West Indies even though they lie outside the main archipelago in the Atlantic Ocean.

Approximately 235,000 square kilometres (91,000 square miles) of wetlands are found on these islands. The wide range in island size, topography and local climatic conditions have created immensely varied wetland types. The most extensive are found in the large islands of the Greater Antilles, the French Antilles and Trinidad, with the largest being the Ciénaga de Zapata in Cuba, which covers 3,400 square kilometres (1,300 square miles). Generally, these are high islands, with much of their central land area over 1,000 metres (3,300 feet). Many support large rivers with floodplain and estuarine environments in which complex wetland systems can develop. In contrast, in low islands, such as Anegada, the Bahamas and the Caicos Islands, where the height of the land rarely exceeds 30 metres (100 feet), there are few rivers and wetlands are restricted to saline coastal types.

Freshwater marshes are rare in the Caribbean. This is because some have been converted into agricultural sites, particularly for rice cultivation, but also because the climate favours swamp forests over herbaceous aquatic plants. Nonetheless, there are some large marshes in the greater Antilles and Trinidad and most of the high eastern Caribbean islands have small patches.

Very few Caribbean wetlands are legally protected or managed for resource use. The larger sites have received the most attention and several of them are reserves or are included in reserves, such as the Zapata National Park, which incorporates the Ciénaga de Zapata. In general, however, there is widespread degradation of mangrove areas by cutting for charcoal burning and coastal pollution, waste dumping and land reclamation, while many freshwater sites are also suffering from encroachment for farming coupled with burning.

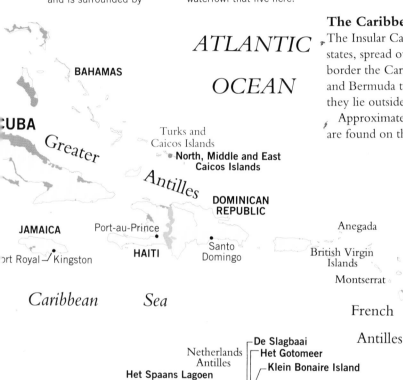

Central America

The Central American isthmus connects the continental masses of North and South America, and separates the world's two largest oceans. In spite of its small size, only 500,000 square kilometres (190,000 square miles), it possesses an extraordinary biological, physical and cultural diversity. Wetlands are an important part of Central America. Freshwater swamps, peatlands, swamp palm forests, floodplains, coastal lagoons, and mangrove forests are common features and support a wide range of uses by local communities.

   The central mountain chain is a strong influence on the climate of the region. It acts as a barrier to wind circulation, and makes the Caribbean much more humid than the Pacific. Close to 70 per cent of the region's total rainfall drains into the Caribbean, where river basins are larger, water volumes higher and river flow fairly regular throughout the year. Rivers on the Pacific side are shorter, with less regular and lower water volumes, and a distinct dry season. On the Pacific coast, mangrove forests and temporary nontidal wetlands are most common; while on the Caribbean side the variety of wetlands is greater, including permanent and temporary freshwater marshes, palm swamps and freshwater swamp forests, forested peatlands, and important coastal lagoons, coral reefs and seagrass beds.

Population pressure

While climate and hydrology explain the distribution of the different types of wetlands in the region, demography and history help explain the pattern of wetland use. During colonial times, between the 1500s and 1700s, population growth was slow; even by the early 1900s only about 3 million people inhabited Central America. By 1990, however, the population had grown to

Left **A banana plantation in Bocas del Toro, Panama. Banana plantations are causing severe damage to much of Central America's Caribbean coast. Banana plants cannot survive periods of waterlogging, and the extensive draining canals that are created to keep the plants alive carry fungicides, fertilizers and pesticides, which are used in high quantities on the plantations, directly into many rivers.**

Left **An area of flooded forest at Bocas del Toro, Panama. Over recent years, the large population growth over much of Central America has brought about increased pressure on vast areas of freshwater wetlands. A major cause of wetland drainage and conversion is the growing numbers of citrus fruit and oil palm plantations.**

Below **A shrimp farm on the west coast of Nicaragua. Shrimp mariculture is presently responsible for part of the widescale destruction of mangrove forest in the region. Yet evidence suggests that replacing mangroves with shrimp ponds is proving to be less profitable than was originally hoped.**

30 million, with 44 per cent less than 15 years old. It is estimated that by the year 2000 there will be between 35 to 40 million Central Americans.

Most of the population is concentrated in the central highlands and along the Pacific coast where there is now severe deforestation, erosion and depletion of natural resources. In recent years, this pattern of resource loss has led to emigration into the extensive and humid Caribbean plains where the new settlers have brought with them their traditional production systems from drier areas. As a result, slash-and-burn agriculture and extensive cattle ranching have led to erosion of the fragile soils. Today, wetland degradation on the Caribbean side continues with the expansion of the agricultural frontier, particularly for banana, citrus fruit and oil palm plantations. Drainage of freshwater wetlands is now common, and siltation, eutrophication and high levels of pesticides are affecting the previously pristine coastal lagoons of the Caribbean coast.

Along the Pacific, where almost all of the dry forests have been lost, mangroves have become an important source of firewood. This is now used for both domestic and industrial purposes, including salt production. While such practices yielded substantial benefits for the population, it is questionable whether mangroves in such areas as the Pacific coast of Nicaragua can sustain these current high levels of exploitation. Shrimp mariculture is also a major threat to the mangroves of the region. As a result of almost negligible land costs – concessions of mangrove land for mariculture in the Gulf of Fonseca in Honduras cost less than US$1 per hectare per year (US$0.40 per acre) for the first three years – most shrimp ponds are being constructed in cleared mangrove forests, even though it has been demonstrated elsewhere that operational costs are lower and production higher in other coastal sites.

These pressures have combined to decrease substantially the area of mangroves. For example, Guatemala lost half of its mangroves between 1960 and 1985. In Panama more than 400 square kilometres (150 square miles) of mangroves have been lost to agriculture, ranching and shrimp mariculture.

Conservation policies

In the face of these pressures, the degree of protection offered to wetlands in Central America differs in each country. Guatemala, Costa Rica and Panama are signatories to the Ramsar Convention, and the Laguna del Tigre Biotope in Guatemala, Palo Verde National Park and Caño Negro Wildlife Refuge in Costa Rica, and Golfo de Montijo in Panama have been designated as Wetlands of International Importance. Meanwhile, Nicaragua has created a unit within the Ministry for Natural Resources (IRENA), that deals directly with wetlands and coastal areas.

Each country in the region has created its own Protected Areas System and important wetlands have been included, such as Crooked Tree Wildlife Sanctuary in Belize, Los Guatusos Wildlife Refuge in Nicaragua, Chocón-Machacas Biotope in Guatemala, and Cuero y Salado Wildlife Refuge in Honduras. Although these wetlands were declared protected areas mainly because of the important wildlife they harbour – crocodiles and caimans, manatees and migratory and resident waterfowl – the close linkages between human populations and wetlands in Central America are now widely realized. If wetland management efforts in the region are to be successful, local populations must be involved and receive benefits from this.

El Salvador's Laguna El Jocotal provides an excellent example of the way in which people benefit from wetland conservation. Lying in a volcanic crater, Jocotal is a freshwater lagoon which varies in size from 5 square kilometres (2 square miles) in the dry season to 15 square kilometres (6 square miles) in the rainy season. More than 130 species of aquatic birds use the lagoon, with the tree duck *Dendrocygna autumnalis* occurring in large numbers. By 1977, hunting and loss of nest sites – a consequence of deforestation around the shores of the lagoon – had combined to reduce the population of this particular species to less than 500 individuals. In response, the Salvador National Park Service began a collaborative programme with the local communities to control hunting and install nest boxes. In return, locals have been allowed to collect a proportion of the eggs from some of the nests, thus obtaining an important protein supplement to their diet. This strategy is helping to restore the tree duck population to its former extent. More important, more than 80 local people have worked directly and indirectly for the project and a number of these have put up their own nesting boxes. This has in turn contributed to a substantial appreciation by the local communities of the importance of Laguna El Jocotal and the benefits that can be obtained from it on a sustainable basis.

Multiple Uses of Mangroves

In many parts of the Pacific coast of Central America as well as in the Caribbean, cutting mangroves to burn and produce charcoal is an important subsistence activity; charcoal is not only used as a home cooking fuel by local people but it is also burnt in barbecues in the urban centres and tourist resorts. To make charcoal, the cut trunks and branches are stacked in a mound and covered with mud or soil and vegetation to ensure that the wood smoulders slowly for several days. The longer the wood smoulders the better the quality of the charcoal.

In Central America, it is common for the larger trees to be used for construction purposes, while thinner poles and branches form the principal source of roofing materials in many coastal areas. In Costa Rica the national telephone company is also investigating the feasibility of using mangrove timber for telephone poles in place of expensive imported softwoods. And in Costa Rica and Nicaragua bark is harvested for the extraction of tannins used in curing leather.

In many areas all of these multiple uses are possible, together with fishery and wildlife harvest. However, this requires careful management in order to avoid overuse of one or more mangrove product. Charcoal cutting in particular can severely degrade mangrove forests, especially if *Rhizophora* is used, as this species does not coppice like *Avicennia* and *Laguncularia*. Where mangrove stands are cleared entirely, the exposed soil dries out quickly and may become too salty for seedlings to grow. Good management preserves mature trees at intervals to provide shade and to aid re-seeding and may involve selective replanting. In St Lucia, charcoal cutters have formed a community association to ensure management of their mangrove charcoal forests, while in Costa Rica a mangrove cooperative, Coopemangle, provides the institutional basis for community management of the mangroves of the Tierra-Sierpe Delta.

Above **Mangroves provide an array of services to local people. As well as providing fuel for cooking and material for building, mangroves also support important fisheries. This service can be maintained in a completely sustainable way and without the need for habitat destruction.**

Right **Charcoal production provides an important source of income for many people in Central America and the Caribbean. The charcoal is used by local people as well as tourists.**

Below **A young southern opossum (*Didelphis marsupialis*)** shows its sensitive whiskers growing from its long snout. The southern opossum is essentially terrestrial, feeding on fruit, insects and small vertebrates; however, during times of flood, this animal becomes arboreal to avoid the waterlogged ground.

Bottom **The marine toad (*Bufo marinus*)** is one of the largest species of toad, growing to a length of 24 cm (10 in). The marine toad feeds on a wide range of prey from small rodents and birds to insects. Glands on the side of the toad's body produce a toxic substance that is powerful enough to kill some of its mammalian predators.

SI-A-PAZ (Yes to Peace)

For much of the 1980s, Central America was known throughout the world as a region of conflict. Nicaragua, El Salvador and Guatemala have all suffered from prolonged fighting, while their neighbours have been used to provide clandestine bases for guerilla armies and their weapons, as well as safe havens for refugees. Today, peace has replaced war in the region, with growing national and international investment being made to heal the wounds of conflict and address the fundamental problems which have led to so much human suffering and environmental damage.

While the reasons for most wars are complex, poverty caused by the degradation of the natural resources upon which most rural people depend has been the major driving force for social unrest throughout Central America. Today, therefore, the challenge for peace in the region lies not just in putting away guns and finding a lasting political solution, but also in pursuing forms of development investment and resource management that do not destroy the environment.

Working for peace

In 1988, Costa Rica and Nicaragua took up this challenge and initiated SI-A-PAZ – which literally translates as "Yes to Peace". This bi-national programme was designed to create a system of protected areas along the San Juan River which forms the border between the two countries. Here, in an area of poor soils with limited agricultural potential, sustainable development means working with local people so that they can draw tangible benefits from the resource. Water, wetlands and the resources they keep alive, are central to this approach.

The freshwater wetlands associated with the Lake of Nicaragua (also known as Lake Cocibolca) and the lagoons, estuaries and marine wetlands of the Caribbean coast, are a dominant feature of the region and are used intensively by local people. In recognition of this importance, SI-A-PAZ is pursuing actions to develop a long-term

conservation plan for the wetlands of the Caño Negro Wildlife Refuge in Costa Rica, and Los Guatusos Wildlife Refuge, which lies across the border on the southern shore of the Lake of Nicaragua. In a similar way, a conservation programme for the wetlands of Tortuguero National Park and Barra del Colorado Wildlife Refuge on the Caribbean coast of Costa Rica, will in due course be linked to efforts to establish conservation areas in the Indio-Maíz Reserve and the delta of the San Juan River in Nicaragua.

These initiatives are designed to conserve the most important natural areas by working with the local communities to develop approaches to resource use which are both sustainable and yield significant benefits. This is being done in the face of growing pressure from commercial interests. Over the last 10 years, tree felling, the expansion of agriculture and extensive cattle ranching have encroached directly on many of the wildlands, resulting in erosion in the upper watershed and increased sedimentation rates in some of the lowland wetlands.

The biggest threat to the wetlands of the Caribbean coast, on the other hand, are the many banana plantations which have been established along the alluvial plains of the most important river systems. Banana plants are very sensitive to waterlogging, and a complex series of drainage canals have been dug in the plantations. These drain directly into the rivers carrying with them many of the fungicides, insecticides and fertilizers which are used heavily in the plantations.

Biodiversity

The fauna of the SI-A-PAZ area is among the richest in Central America. In Tortuguero alone, 405 species of bird, 184 fish and 97 mammals have been recorded. The mammals include a number of endangered species, such as the manatee (*Trichechus manatus*), the jaguar (*Felis onca*), the puma (*Felis concolor*), the tapir (*Tapirus terrestris*) and four species of monkey. Resident and migratory birds are also numerous, among them are found great white herons (*Casmerodius albus*), wood storks (*Mycteria americana*), white ibis (*Eudocimus albus*) and snowy egrets (*Egretta thula*).

Top left **Contra soldiers in northern Nicaragua. After almost a decade of conflict during the 1980s, Central America is now at peace. The SI-A-PAZ (Yes to Peace) programme between Costa Rica and Nicaragua is an important effort to preserve the natural resources of the region and ensure that the local people are able to benefit from the land. In the past, the loss of these resources has added to the poverty of many rural people and helped to fuel civil unrest.**

Above **The forests and wetlands of the Atlantic coast of Nicaragua now have a better chance of surviving unscathed since hostilities in the region ceased. However, the battle against commercial pressures continues.**

Far left **A young South American tapir searches the ground with its sensitive snout for its diet of leaves, buds, fruits and grasses. The South American tapir is nearly always found near water, and is a good swimmer.**

As well as the numerous fish, the rich aquatic fauna includes several species of freshwater turtle, such as *Chrysemis ornata*, and sea turtle (*Chelonia mydas* and *Dermochelis coriacea*); and crocodilians such as *Caiman crocodilus* and *Crocodylus acutus*. These populations were once heavily depleted but have now recovered substantially, particularly after the establishment of Caño Negro as a Wildlife Refuge in 1984.

Fish populations are abundant in the SI-A-PAZ area. Some are of interest for sport fisheries, such as the tarpon (*Megalops atlanticus*), while others such as the relict fish *Atractosteus tropicus* and the freshwater bullsharks (*Cacharhinus leucas*) of the Lake of Nicaragua have spurred interest in the scientific community. Different species of cichlid fish (*Cichlasoma* spp.) and *Rhamdia guatemalensis* are the base of subsistence fisheries.

As more and more of the freshwater wetlands on the Pacific coast of Nicaragua and Costa Rica are being affected by drainage and water diversion, the biological role of the wetlands associated with SI-A-PAZ has become increasingly significant. In recognition of this, Costa Rica listed Caño Negro as a Ramsar Site in 1991 and Tortuguero National Park and Barra del Colorado Wildlife Refuge will be included in the near future.

Northern South America and the Amazon Basin

The northern half of South America is dominated by the Amazon drainage basin. Covering a total of some 7 million square kilometres (2.7 million square miles) the basin contains some of the most extensive floodplain forests, known as *várzea* forest, in the world. The rivers which unite to form the River Amazon flow from headwaters in Bolivia, Peru, Ecuador, Venezuela and Brazil. The largest is the Solimoes, which, after joining the Río Negro east of Manaus, northern Brazil, forms the middle Amazon and flows over 1,200 kilometres (720 miles) to the sea. At its mouth, near Belém, on the north coast of Brazil, the Amazon releases over a sixth of the freshwater carried by all the world's rivers.

The mighty Amazon

Along its course, the Amazon, and its tributaries, inundate 50,000–60,000 square kilometres (19,000–23,000 square miles) of the middle and lower Amazon for several months each year. The deepest parts of this vast forested floodplain are occupied by lagoons which expand and shrink with the flood. The most extensive lagoons cover over 2,000 square kilometres (770 square miles). In the nutrient-poor rivers there are numerous palm-shrouded channels, or *igapos*. At the mouth of the river, the tidal influence extends as far as 1,000 kilometres (600 miles) inland, creating vast areas of seasonal and tidal floodplain. The biggest island is Marajo, which divides the estuary into two branches. The estuary includes areas of flooded grassland where only the *tesos*, mounds built by pre-Columbian Indians, emerge above the water.

Moving east of the Amazon, the coast of Para and Marañhao is indented with more than 35 major inlets and estuaries, each fringed with mangrove swamps. Inland, there are fresh to brackish lagoons and marshes, riverine marshes, areas of seasonally flooded grassland, palm groves and patches of forest.

North of the Amazon Basin lies the Orinoco River, which rises on the Guyana Shield. Here, the river shares several tributaries with the Río Negro, the most famous being the Casiquiare. Northeast of this region, the Caroni River rises in the Gran Sabana, joining the Orinoco some 200 kilometres (120 miles) before it reaches the sea.

In the highlands of Colombia and Ecuador, along the upper courses of the Río Magdalena, there are bogs, marshes and oxbow lakes which are important for resident and migratory waterfowl. And in the High Andes there are many glacial lakes and mountain rivers, many of which are important habitats for waterfowl species, some of which are endemic.

Coastal wetlands

On the Caribbean coast of Colombia, the Ciénaga Grande de Santa Marta forms a complex of wetland habitats near the mouth of the Río Magdalena.

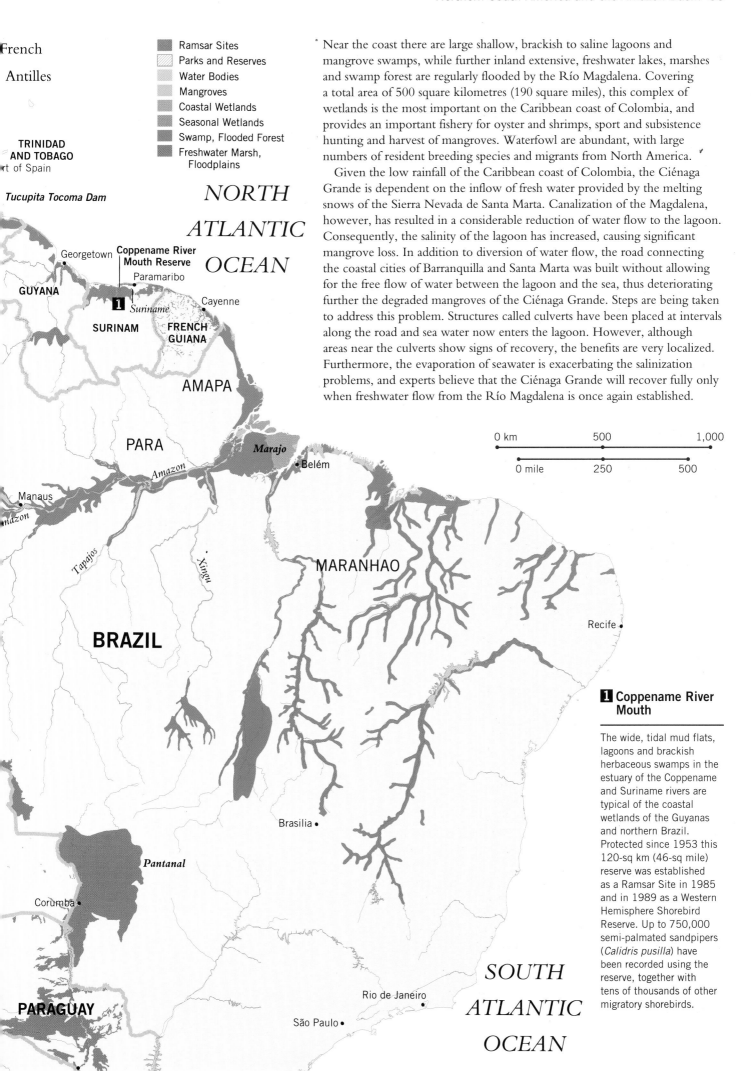

French

Antilles

**TRINIDAD
AND TOBAGO**
rt of Spain

Tucupita Tocoma Dam

Ramsar Sites
Parks and Reserves
Water Bodies
Mangroves
Coastal Wetlands
Seasonal Wetlands
Swamp, Flooded Forest
Freshwater Marsh,
Floodplains

NORTH

ATLANTIC

OCEAN

Georgetown   **Coppename River
Mouth Reserve**

GUYANA   Paramaribo

Cayenne

1 *Suriname*

SURINAM   **FRENCH
GUIANA**

AMAPA

PARA

Marajo

Amazon   • Belém

Manaus

mazon

Tapajos

Xingu

MARANHAO

BRAZIL

Recife •

Brasilia •

Pantanal

Corumba •

PARAGUAY

Rio de Janeiro
•

São Paulo •

SOUTH

ATLANTIC

OCEAN

Near the coast there are large shallow, brackish to saline lagoons and mangrove swamps, while further inland extensive, freshwater lakes, marshes and swamp forest are regularly flooded by the Río Magdalena. Covering a total area of 500 square kilometres (190 square miles), this complex of wetlands is the most important on the Caribbean coast of Colombia, and provides an important fishery for oyster and shrimps, sport and subsistence hunting and harvest of mangroves. Waterfowl are abundant, with large numbers of resident breeding species and migrants from North America.

Given the low rainfall of the Caribbean coast of Colombia, the Ciénaga Grande is dependent on the inflow of fresh water provided by the melting snows of the Sierra Nevada de Santa Marta. Canalization of the Magdalena, however, has resulted in a considerable reduction of water flow to the lagoon. Consequently, the salinity of the lagoon has increased, causing significant mangrove loss. In addition to diversion of water flow, the road connecting the coastal cities of Barranquilla and Santa Marta was built without allowing for the free flow of water between the lagoon and the sea, thus deteriorating further the degraded mangroves of the Ciénaga Grande. Steps are being taken to address this problem. Structures called culverts have been placed at intervals along the road and sea water now enters the lagoon. However, although areas near the culverts show signs of recovery, the benefits are very localized. Furthermore, the evaporation of seawater is exacerbating the salinization problems, and experts believe that the Ciénaga Grande will recover fully only when freshwater flow from the Río Magdalena is once again established.

0 km	500	1,000

0 mile	250	500

**1 Coppename River
Mouth**

The wide, tidal mud flats, lagoons and brackish herbaceous swamps in the estuary of the Coppename and Suriname rivers are typical of the coastal wetlands of the Guyanas and northern Brazil. Protected since 1953 this 120-sq km (46-sq mile) reserve was established as a Ramsar Site in 1985 and in 1989 as a Western Hemisphere Shorebird Reserve. Up to 750,000 semi-palmated sandpipers (*Calidris pusilla*) have been recorded using the reserve, together with tens of thousands of other migratory shorebirds.

The Llanos

In a belt running from the Cordillera de Merida in the west of Venezuela to the Guyana Shield and the Orinoco River to the south and east, lie the Venezuelan Llanos, part of the floodplain of the Apure and Orinoco rivers and their tributaries. The Llanos are among the largest wetland areas in South America, covering over 100,000 square kilometres (40,000 square miles) in Venezuela alone (they also extend across the border into Colombia). This vast wetland region forms a mosaic of slow-flowing rivers and streams, oxbow lakes, riverine marshes and swamp forest, permanent and seasonal freshwater lakes, ponds and marshes, and large areas of seasonally inundated grassland and palm savanna. For six months of the year, the Llanos become an ocean of water and grass, turning, as the floodwaters recede, to a dry, dusty plain.

The cycle of flood and drought which characterizes the Llanos determines the seasonal cycle of the fauna and flora of the floodplain. The summer dry season is a time of reduced growth, when annual plants die off and the smoke of frequent forest fires stretches across the dry savanna. During this time, most animals move to the valley bottoms and along the rivers, where the land remains wet. With the winter rains comes the flood which fills the rivers and lakes, and stretches across the plain. This stimulates the germination of annual plants and the sprouting of perennials, nourishing the region's food chains.

A wealth of life

The floodplain supports a rich and varied flora and fauna. The vast expanses of aquatic grasses are dotted with palms (*Mauritia* spp. and *Copernicia tectorum*), which are used by the locals for roofing material. In general, however, the rigours of the drought and fires limit the development of more complex plant communities. Instead, well-developed woodlands, for example, tend to establish along river beds, where they are protected from fire.

Below **The Apure and Orinoco rivers and their tributaries meander through the savanna lands of the Llanos in Venezuela. Much of the woodland is concentrated around the margins of the rivers where water is found throughout the year. A large part of the Llanos is privately owned and used for cattle ranching, although some protected areas have been established. Among the largest is the Aguaro-Guariquito National Park.**

Top right **Tree frogs (*Hyla* sp.) are abundant in the open woodland of the Llanos. Most species are nocturnal, and it is only at night that they demonstrate their superb adaptation to life in the trees. A combination of suckered toes and a remarkable sense of balance makes these amphibians exceptionally agile animals.**

Centre right **The common iguana (*Iguana iguana*), like the tree frogs, is found in great numbers in the Llanos woodland. The iguana is mostly arboreal, its favourite haunt being branches that overhang water. If disturbed, however, the reptile drops into the water, seeking safety from its predators below the surface. Mature adults reach up to 1.8 m (6 ft) in length. Common iguana meat is highly prized by local peoples.**

Bottom right **The red-bellied piranha reaches a maximum length of less than 30 cm (12 in), yet the piranha has gained the reputation as one of the most feared fishes in the world. Piranha are found in most South American rivers and usually occur in large shoals that contain hundreds, even thousands, of these fiercely carnivorous fish.**

The Llanos' rivers are rich in fish, many of which are caught and eaten. Other fish, such as the stingray (*Dasyatis americana*) and the electric eel (*Electrophorus electricus*), can cause painful wounds and are feared by the local population. Most famous of all the fish, however, is the caribe or red piranha (*Serrasalmus nattereri*), which has developed an international reputation. When local people have to swim across rivers infested with piranhas, they fear them much more than the local crocodilian, the caiman.

The caiman (*Caiman crocodilus*) is one of the most important residents of the Llanos. Once described as being so numerous as to form a "moving pavement" in the water, poaching has reduced numbers substantially. Fortunately, conservation efforts are increasing and the outlook is much more positive than it was in the late 1970s. A major factor in conservation has been the controlled harvest for the caimans' skins and meat, which has given local people an incentive to protect the species.

The Llanos also provide one of the most important habitats in the world for wading birds. Over 100 breeding colonies of herons, storks and ibises have been recorded in a single year, with the largest holding 32,000 pairs. In 1983, 65,000 pairs of scarlet ibis (*Eudocimus ruber*) were located in 22 colonies, as well as over 5,500 wood storks (*Mycteria americana*) and 185 jabiru (*Jabiru mycteria*).

The grasslands support a rich variety of herbivores, most noticeably the capybara (*Hydrochoerus hydrochaeris*) and the white-tailed deer (*Odocoileus virginianus*). Today, the deer mix with the cattle, which combined total 10 million individuals grazing on the floodplain, and together they have come to characterize the Llanos in the minds of many people. At the top of the food chain is the jaguar (*Panthera onca*), which is affectionately known as "tio tigre" (uncle tiger) by the local people.

Peoples of the Llanos

Prior to European colonization of the Americas, the Llanos were inhabited by different tribes over the centuries, stretching as far back as 15,000 years ago. These tribes were hunter-gatherers who lived off the rich resources of fish, wildlife and wild plants. Today, less than 10,000 indigenous people remain.

European settlement of the Llanos did not begin until relatively recently. The first explorers saw it as empty, without the gold and other easy riches that they sought. Only in the late 1700s and early 1800s did efforts at colonization lead to the beginnings of a long-term presence based upon cattle and centred around missions. Even today the Llanos have a low population density. The States of Apure and Guarico, which include almost all of the lowest part of the Llanos, cover 14 per cent of Venezuela, but with 550,000 people hold only 3.8 per cent of the population.

The past 200 years have left their mark on the Llanos. The most significant change is the progressive impoverishment of the region's soils under continuous grazing, the loss of several species through fires, and the slow extermination of others through illegal hunting. However, there is today a substantial increase in conservation investment in the Llanos, with government and non-governmental organizations, and private landowners, working together to establish forms of land use that can yield economic benefits without destroying the natural riches of the region.

The Orinoco Delta

The Orinoco Delta covers 36,500 square kilometres (14,000 square miles). It lies in Venezuela's Delta Amacuro Federal Territory and includes the coastal swamps of Monagas and Sucre States. Seventy thousand people live in the Federal Territory, of which 70 per cent are concentrated in and around the capital Tucupita. Their main economic activities are agriculture, in particular growing cacao for cocoa production, and cattle ranching. The cattle are shipped out from the delta's savanna islands during the season of high waters.

The indigenous population consists of some 16,000 members of the Warao Tribe, who practise subsistence agriculture and fishing. Salted fish, especially morocoto (*Colossoma brachypomum*), are exported to the major population centres that border the region. Some small sawmills are in operation in the interior of the delta, and there are three palm heart canneries. The economically valuable palm heart extraction is based on a sustainable government management plan whereby sufficient shoots are left for the plants to regenerate.

Wildlife found in the delta includes the spectacled caiman (*Caiman crocodilus*), which was subjected to intense commercial hunting during the early 1970s. However, a survey in 1988 indicated that the populations of this species had recovered. This is not true, however, of the Orinoco crocodile (*Crocodylus intermedius*), which was almost wiped out during the 1960s, and now only scattered individuals survive in the delta. Likewise, little or no recovery has been seen with the herbivorous manatee (*Trichechus inunguis*) nor with the giant otter (*Pteronura brasiliensis*).

The Man-made Wetlands of Guyana

The coast of Guyana is bordered by a plain about 25 to 35 kilometres (15 to 21 miles) wide, most of which lies below the level of mean high tide. For over three centuries this coastal plain has been the focus of development investment. As early as 1621, the famous trading company the Dutch West India Company was formed and the Essequibo colony established. Then between 1655 and 1680, construction of a sea wall commenced, the purpose of which was to reclaim land from both the sea and the freshwater swamps. An interlaced network of drainage channels was built, and at the seaward end of these channels kokers were installed, gate-like devices that allow the land to be drained at low tide. Plantations were laid out in strips running inland, and a dam and a network of irrigation canals were built to provide water for farmlands and populated areas.

Today, more than 90 per cent of Guyana's estimated population of 890,000 lives on the coastal plain, and investment in drainage for agriculture and other development projects has continued. Over the past 40 years, the sea defence, and drainage and irrigation system have been expanded through several major land development projects.

The cost of agriculture

The net result of three centuries of modification of the coastal plain is that the region today forms one of the largest man-made wetlands in the world. More than 2,000 square kilometres (770 square miles) of land within the coastal strip are irrigated and drained by 14,000 kilometres (8,400 miles) of man-made channels. Landward, the artificial wetlands and their associated swamps cover 5,400 square kilometres (2,100 square miles). However, in common with the rest of the world, Guyana's wildlife has suffered from the conflict between agriculture and wildlife conservation. From 1967 to 1972, misuse of pesticides and herbicides is believed to have killed large numbers of fish, hatchling caiman and other aquatic wildlife. During the same period, and again between 1984 and 1988, spectacled caiman populations were depleted through commercial hunting.

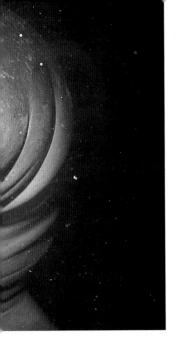

Caroni River Development

Left **The Amazonian manatee is the smallest of the manatees, and is an almost exclusively freshwater species. The main cause for the drastic drop in numbers in the Orinoco Delta is hunting for skin, oil and flesh, although in Brazil the manatee is thought to be recovering from decades of persecution.**

The Caroni Hydroelectric Development Project is one of the success stories of hydroelectricity. The project consists of the Guri Dam, which regulates the flow of the Caroni River and a further three dams, Tocoma, Caruachi and Macagua II. The latter are being constructed downstream in the form of a "staircase" of man-made lakes. Construction of the first stage of the Guri began in 1963. At the end of 1986, the lake was brought up to its final elevation of 270 metres (890 feet) above sea level. Macagua II is currently under construction. When completed the project will produce 17,980 megawatts (a million watts). The lower Caroni is unusual in that instead of forming a broad alluvial plain it flows through a relatively narrow and deeply cut channel. The impacts of flooding this region, therefore, have occurred on terrestrial ecosystems, which, in this area, are less fragile and less important to wildlife than wetland habitats. Damming in 1968 possibly cut the migration routes of some fish species, but there is no evidence that there was ever significant fishing activities in the Caroni prior to that date. The Guri empoundment has 134 cubic kilometres (32 cubic miles) of useful storage, which permits the total regulation of the flow of the Caroni River. The Caroni contributes 14 per cent of the Orinoco's mean annual flow.

Wetland creation

The hydrological impact of regulating the Caroni will be a slightly higher water level during the dry season. It is doubtful that this will affect the ecosystems of the Orinoco Delta. The Guri Lake covers 4,300 square kilometres (1,660 square miles). In general, the lake has retained the original characteristics of the Caroni River, and flooding by stages apparently reduced the possibility of excessive eutrophication. The three lakes that will be produced by the dams downstream will be smaller. Tocoma will cover 9 square kilometres (3.5 square miles), Caruachi, 25.5 square kilometres (10 square miles) and Macagua II, 47 square kilometres (18 square miles). However, since the operating levels of these lakes should not vary by more than about a metre (3 feet), it is expected that they will develop stable wetland ecosystems along their shallow shorelines.

Left **Installing generators in the powerhouse of the Guri Dam, the main feature of the Caroni Hydroelectric Development Project. A series of four lakes and dams will eventually regulate the flow of the Caroni, one of the major tributaries of the Orinoco. Despite this large-scale interference with the natural flow of the river, the wetlands of the region are unlikely to be seriously affected by the project.**

The Amazon

Since its encounter with the first Europeans in 1542, the River Amazon has fascinated successive generations of modern society. Coming only 10 years after the Incas of western Peru had been overwhelmed by Francisco Pizarro, the discovery of this mighty river led to successive expeditions in search of further riches. The Europeans knew of the societies which had flourished along the Nile, the Tigris-Euphrates, the Ganges and the Indus, and sought similar wealth along this, the largest of the world's rivers. That they found only Indian tribes living from the resources of the forest and river led to disillusionment, and when in 1639 Pedro Teixeira travelled 2,000 miles upstream and claimed for Portugal all of the land which now lies east of Ecuador, Spain saw this as being of little importance.

Over three centuries later, the Amazon Basin remains one of the world's great wildernesses; a river of awesome power bordered by the Earth's largest expanse of tropical forest. Today, the Amazon is South America's new frontier, an area rich in mineral resources where many hope to find solutions to problems of widespread poverty, rising population, inflation and external debt. As the forest has fallen, concern for the future of the Amazon region and the people who live there has received growing international concern, giving rise to claims of foreign interference in national affairs. To date, most of the debate over the Amazon has focused upon the forest resources with little consideration given to the river and its associated lakes and floodplains, arguably the most productive but fragile parts of the system.

A stretched resource

Today, fishing is the most important economic activity along the rivers and in the adjacent *várzea* floodplain forest and lakes. Fish provide over 60 per cent of protein consumed by the 100,000 families that live along the river banks in the States of Amazonas, Para and Amapa. Their traditional small-scale fishing, however, is now in conflict with larger-scale fishing from the ports of Manaus and Belém. As the rural poor have migrated to these urban centres in search of work, demand for the cheap fish protein has grown substantially. Commercial fishing has intensified in response to this demand, and also to meet the needs of the larger towns of the south. For this reason, today's commercial fishing has reduced fish sizes and the productivity of the fisheries. This has exacerbated the problems of the rural communities.

Because terrestrial sources of food such as cassava provide little protein, river communities have relied heavily on fishing to provide essential dietary protein. For example, 61 per cent of the animal protein eaten by rural people in the Ucayali Valley of Peru comes from fish. In an effort to protect the floodplain fisheries and the people dependent upon them, the past few years have seen a number of initiatives to establish fishery reserves in the floodplain lakes. The community leaders urging the establishment of these reserves, however, have had to confront powerful commercial interests currently overfishing the systems. Already one prominent leader has been killed.

Consequences and action

The development boom in the Amazon Basin has led to widespread deforestation, resulting in soil erosion, river siltation and declining productivity of many of the aquatic ecosystems. Meanwhile, the expansion

Top **The Tocurui Dam on the Tocantins River, Brazil. The Tocurui is the largest dam ever built in a rain forest region, and its environmental impact is still being assessed. The photograph was taken three days after the dam became operational. Behind it, some low-lying vegetation has already become flooded. Downstream, it is thought the dam will reduce water flow and alter rates of sedimentation.**

Above **Riverside village in Brazil. Throughout the region, traditional patterns of human activity persist, and wetlands provide food for a scattered population.**

Right **Fish market in Belém, Brazil. Traditional methods ensured fishing was sustainable, but recently, the introduction of large-scale commercial fishing has aroused fears that some rivers will soon become overfished.**

of gold mining in many of the southern tributaries has resulted in severe mercury contamination. Many areas of várzea have also been cleared for livestock and commercial crops such as jute, maize and rice. While it is hoped that such agricultural practices will benefit from the fertile soils of the floodplain and generate sustainable high yields, it is at the expense of traditional river communities, who must find other land or move to the city. Clearance of the floodplain is believed to have led to a 23 per cent decline in Amazon fish catches between 1970 and 1975.

Today, most international attention is focused upon the dams of the Amazon Basin. So far three have reached reservoir stage, and there are plans for a further 76 to be built. But while increased supplies of clean power are essential for effective and environmentally sound exploitation of the region's mineral wealth, the completed dams have been plagued with cost overruns, environmental and social impacts and doubts over their long-term viability.

The development pressures upon the Amazon are enormous. A substantial part of the basin's natural wetlands have already been destroyed by human activity. However, it is not too late for action to address these. Legislation and effective policing procedures are needed to stop mercury pollution from gold mining; proposals for new dams must consider the total impact on the region and its people, and systems of integrated resource use on the floodplain need to be developed and promoted as part of government policy.

Southern South America

From the high-altitude lakes of the Andes, through the vast floodplains of the Paraná-Paraguay river basin to the mangroves and lagoons of the Brazilian coast, the southern cone of South America supports some of the most varied wetlands in the world. Although these wetlands have supported human populations for thousands of years, today the southern cone is the site of growing conflict over conservation and development policy. The nations of the region are seeking to maximize economic output and raise GNP in the face of rising population and mammoth external debt. In support of this, a substantial increase in investment in wetland conservation is needed in an effort to ensure that the major wetland benefits are maintained, and that the people of the region can continue to gain from these resources.

The Paraná

Stretching 4,000 kilometres (2,500 miles) from the interior of Brazil to its mouth at Buenos Aires, the Paraná River is the second largest in South America; its basin drains 2.8 million square kilometres (1.1 million square miles). Just north of the Argentine cities of Resistencia and Corrientes on the border with Paraguay, the Paraná is joined by the Paraguay River, which also flows from Brazil's interior, for 2,250 kilometres (1,350 miles). The vast region surrounding these river courses is predominantly flat, has a benign climate and possesses some of the most fertile soils on the continent, factors which have combined to make this the most densely populated and industrialized region of South America. The two largest cities of South America, São Paulo and Buenos Aires, lie within the Paraná-Paraguay Basin.

Matching this economic importance, the floodplains of the basin are among the most important in the world. For most of the year, the Pantanal covers over 140,000 square kilometres (54,000 square miles), abutting the borders of Brazil, Paraguay and Bolivia. During the flood season, the Pantanal spreads out to flood 250,000 square kilometres (96,500 square miles) – over 6 times the size of Belgium. For this reason the Pantanal is widely regarded as the largest wetland in the world. To the south, the basin's floodplains continue in virtually uninterrupted progression. They spread through the wet prairie-like Chaco region of Paraguay and Argentina, join the floodplain of the Paraná River south of the confluence with the Paraguay, and terminate in the delta of the Paraná River to the west of Buenos Aires.

The annual flood cycle of the river is critical for the livestock industry which has developed along the river. During the periods of receding flood, small lakes and areas of extensive pasture remain, providing a rich food source for large herds of cattle. During the flood the cattle are removed to higher land and the floodplains are then the domaine of the fish which migrate upstream to breed. Young fish normally colonize the lakes of the floodplain, where they remain for one or two years before returning to the river with the flood.

Bañados del Este

The Bañados del Este, the coastal lagoons and freshwater marshes on the shores of Laguna Merin, cover 2,000 square kilometres (770 square miles) and are among the most important natural ecosystems in Uruguay. In 1984, the Bañados were listed under the Ramsar Convention when Uruguay signed the treaty. They support a high vertebrate and invertebrate diversity, including 35 per cent of the freshwater fish species, 47 per cent of amphibians, 58 per cent of reptiles, 42 per cent of birds and 51 per cent of the mammals of Uruguay. One of the characteristic landscapes of the Bañados del Este are the groves of butia palm, which cover a total of 700 square kilometres (270 square miles). These are now threatened by cattle whose grazing prevents the growth of young trees. Improved protection measures are essential if this unique landscape is to survive into the 21st century.

■ The Upper Paraná Basin

Rapid deforestation over much of the upper basin of the Paraná River has increased erosion and the risk of floods and droughts. The catastrophic floods of 1983 were exacerbated by deforestation and other human intervention in the hydrological capacity of the catchment. Damage was estimated at US$1 billion and an area of some 100,000 sq km (40,000 sq miles) was inundated.

2 Rio Grande do Sul

The coastal plain of Rio Grande do Sul in southern Brazil supports a diverse assemblage of lakes and lagoons of various sizes which lie behind a chain of barrier islands and dune systems. The largest lagoon in Brazil is the Lagoa dos Patos, which stretches 250 km (150 miles) and covers 10,360 sq km (4,000 sq miles). Increasing population and industrial development have increased pressure upon the coastal areas of Río Grande do Sul and industrial and domestic effluents have degraded the lagoons. In the surrounding marshes drainage for pastureland and rice cultivation has given rise to calls for reserves to be established.

Chilean Fjordland

Stretching over 1,500 km (900 miles) from the Gulf of Guafo in the north past the straits of Magellan and Tierra del Fuego to Cape Horn in the south, the fjord coastline and associated offshore islands of southern Chile support a mosaic of sandy beaches and small estuaries with intertidal mud flats and salt marshes. Inland, there are numerous freshwater lakes, marshes, peat bogs, fast-flowing rivers and streams, wet meadows, areas of seasonally flooded grassland and some swamp forest. On the higher ground and in the south there are extensive areas of tundra and bogs fed by melting glaciers and snow.

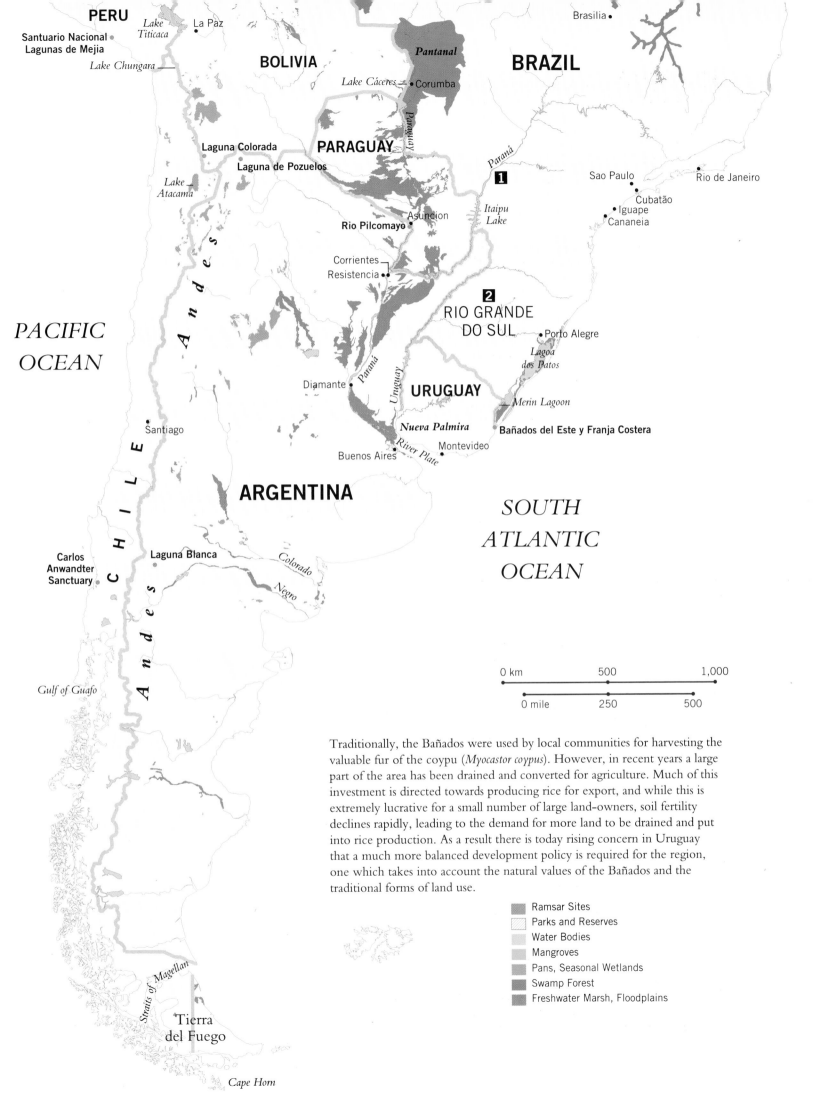

PERU

Santuario Nacional
Lagunas de Mejia

Lake
Titicaca ● La Paz

Lake Chungara

BOLIVIA

Pantanal

BRAZIL

Brasilia ●

Lake Cáceres ● Corumba

Laguna Colorada

Laguna de Pozuelos

PARAGUAY

Lake
Atacama

Asuncion

Rio Pilcomayo ●

Paraná

1

Itaipu
Lake

Sao Paulo ●

Rio de Janeiro ●

Cubatão
● Iguape
Cananeia

Corrientes
Resistencia ●

2

RIO GRANDE
DO SUL

Porto Alegre ●

Lagoa
dos Patos

Diamante ●

Paraná

Uruguay

URUGUAY

Merin Lagoon

PACIFIC
OCEAN

Nueva Palmira

Montevideo

Bañados del Este y Franja Costera

Buenos Aires ●

River Plate

Santiago ●

ARGENTINA

SOUTH
ATLANTIC
OCEAN

Carlos
Anwandter
Sanctuary

Laguna Blanca

Colorado

Negro

C H I L E

A n d e s

Gulf of Guafo

0 km   500   1,000

0 mile   250   500

Traditionally, the Bañados were used by local communities for harvesting the
valuable fur of the coypu (*Myocastor coypus*). However, in recent years a large
part of the area has been drained and converted for agriculture. Much of this
investment is directed towards producing rice for export, and while this is
extremely lucrative for a small number of large land-owners, soil fertility
declines rapidly, leading to the demand for more land to be drained and put
into rice production. As a result there is today rising concern in Uruguay
that a much more balanced development policy is required for the region,
one which takes into account the natural values of the Bañados and the
traditional forms of land use.

Straits of Magellan

Tierra
del Fuego

Cape Horn

- Ramsar Sites
- Parks and Reserves
- Water Bodies
- Mangroves
- Pans, Seasonal Wetlands
- Swamp Forest
- Freshwater Marsh, Floodplains

Wetlands of the Puna

The cold, arid Andean altiplano, known as the Puna, is characterized by extensive tablelands that lie between 3,200 and 4,500 metres (10,500 and 14,800 feet) above sea level. Here there is a diversity of lakes of various sizes and salinities, together with temporary lagoons, freshwater marshes known as bofedales, and bogs. Of the lakes, Lake Titicaca is the most famous. According to legend the sun gods created Manco Capac and his wife Mama Ocllo, the first of the Incas, on Lake Titicaca's Isla del Sol (Island of the Sun). From there they spread out to colonize an immense empire and establish one of the most advanced civilizations of the continent.

The highest navigable lake in the world and the largest in South America, Lake Titicaca was held sacred by the Incas, who prospered on its shores. Prior to the arrival of the Europeans, the lake supported a healthy economy based principally upon rearing llamas and alpacas, as well as fishing and commerce. Today, the lake is shared between Peru and Bolivia, and a sizeable population continues to live on its shores and depend upon its fringing wetlands. The domestic herds are now composed of vicuñas and cattle. The submerged aquatic vegetation, "yacco", is collected as cattle feed. In addition, a type of reed (*Scirpus* sp.) is used in handicrafts and as food. The quantity of fish is estimated at 80,000 tonnes, of which some 5,000–6,000 tonnes are caught annually for direct consumption and sale.

Icy lakes

As well as Lake Titicaca, the other lakes of the Puna, some of which contain icebergs because of the altitude, are home to many endemic species and spectacular communities of waterbirds. Lake Junín in Peru, with endemic species of grebe (*Podiceps taczanowskii*) and frog (*Batracophrynus macrostomus*), is one of the most famous and best studied of these lakes. Three species of flamingo live in the Puna, two of which are endemic to the region. Of these the Andean flamingo (*Phoenicoparrus andinus*) is the biggest and rarest, and Lake Atacama in Chile is its only permanent breeding site. James' flamingo (*Phoenicoparrus jamesi*) is the smallest and nests mainly in the Laguna Colorado in Bolivia, while the third species, the Chilean flamingo (*Phoenicopterus chilensis*), is more widely distributed throughout the south of the continent. It is thought that the flamingos, which feed by filtering phytoplankton and zooplankton, are confined to salt lakes where there are no fish. The most likely reason for this is that in other lakes, fish, which tend to be more efficient feeders, consume the greater part of the available food supply.

Despite the high altitude and relative remoteness of the Puna's wetlands they face a diversity of pressures. In Lake Titicaca the introduction of trout has resulted in a decline in native species, while indiscriminate burning of the reed beds, hunting and grazing all threaten to degrade the ecosystem upon which both people and wildlife depend. Many of the region's lakes are contaminated by run-off from the many precious-metal mines. A third of Lake Junín, for example, is contaminated by residues from the lead, copper and zinc mines of the Cerro Pasco-La Oroya. The basin of Peru's Mantaro River down to the Tablachaca reservoir contains 18 mines, which contaminate the river with toxic residues. This poses a serious problem not only for the river system, but also for the reservoir, which is a future source of drinking water for the 7 million inhabitants of Lima.

Conservation strategies

However, despite these problems progress has been made in recent years to protect the Puna's invaluable wetlands. A concentrated effort by national and international conservation organizations has halted plans to use water from Lake Chungara in Chile for irrigation; parts of lakes Titicaca and Junín are now protected areas, Laguna Colorada in Bolivia is listed under the Ramsar Convention, and the Laguna de Pozuelos in Argentina, which in 1992 was listed as the country's first Ramsar Site, is now a Biosphere Reserve.

Left **Sheep grazing on the Peruvian shores of Lake Titicaca. Many of the region's lakes and wetlands are saline and provide only limited opportunities for pastoralism. Titicaca is freshwater, and the seasonal flooding supports a rich natural grassland suitable for arable farming. Furrows on the surface show the field has recently been used for crops.**

Top **Native llamas share the Titicaca grasslands with introduced sheep. The domesticated llama (*Llama guanicoe glama*) is larger than the wild guanaco (*Llama guanicoe*) from which it was bred thousands of years ago. Although llamas are now raised mainly for wool, they were once important beasts of burden, prized for being able to carry heavy loads.**

Above **Reeds dominate the lives of those who make their homes on Titicaca's "floating islands" – patches of matted vegetation that provide a semi-solid surface, surrounded by water. Huge beds of emergent club rushes extend up to 12 km (7 miles) from the shore in parts of the lake. Harvested, dried, bundled and woven, the reeds provide the island-dwellers with a flexible construction material that can be transformed into homes and boats, as well as a wide range of furnishings.**

The Pantanal

Seen from the air, the Pantanal resembles a patchwork quilt; floodplain lakes and marshes interspersed with one of the most diverse tree floras on the continent. Areas of higher ground provide important refuges for many species of birds and mammals, as well as serving as nesting sites for caiman, and greatly enhancing the richness of the flora and fauna of this vast marshland.

On the ground, the Pantanal supports some of the largest concentrations of waterbirds in Latin America and, together with the Llanos of Venezuela, provides a critical stronghold for the Jabiru stork (*Jabiru mycteria*), the largest of the region's wading birds. The density and diversity of mammals is, however, limited by the incidence of extreme floods. Five times in the 1900s – in 1905, 1920, 1932, 1959 and 1974 – exceptionally high floods inundated large parts of the Pantanal, and even by the early 1980s, the water had still not returned to normal levels. Small terrestrial mammals are therefore absent from the flooded areas, except for those which managed to find refuge on high ground. Intensive cattle management has also converted large areas of the forest habitat to pasture, thus changing a complex habitat to a simple one, and at the same time eliminating most forest mammals.

Home of the jaguar

Although the Pantanal's animal population may be limited in its diversity, it remains one of the most spectacular in Latin America. Despite continued poaching, the Pantanal is one of the remaining strongholds for the jaguar, with a population estimated at 3,500 in 1986. Each adult male ranges over an area of 50–80 square kilometres (30–50 square miles) with females occupying half of this. Both males and females feed on a wide variety of animals from coati (*Nasua nasua*) and caiman (*Caiman yacare*) to tapir (*Tapirus terrestris*). Cattle are believed to be the main prey, although kills account for only a small percentage of the animals dying annually. In the Pocone district, which contains the largest remaining jaguar population in the Pantanal, drowning, disease and starvation reduced the cattle population from 700,000 before the 1974 flood to 180,000 in the mid-1980s.

The main wild prey of the jaguar is the capybara (*Hydrochoerus hydrochaeris*), although populations of this large rodent were also reduced severely following the 1974 flood. Deprived of their favoured habitat on the floodplain, the capybara crowded onto the higher ground, where they were vulnerable to diseases. The marsh deer (*Blastocerus dichotomus*), the largest deer in South America, has also been affected by the flood. In 1977, the population was estimated as 5,000–6,000, but reproduction has been poor and disease has reduced numbers.

Skin trade

The Pantanal resident that has suffered most from poaching is the caiman. Groups of professional hunters move into the area during the dry season and operate on private lands with or without the landowners' permission. An estimated one million hides are smuggled out of Brazil each year, being taken directly, or indirectly through Bolivia, to Asunción in Paraguay. In the face of this pressure,

Left **A caiman at rest among aquatic vegetation. There are two species of caiman found in the Pantanal – spectacled and broad-nosed. Both are much smaller than the black caimans found in the Amazon.**

Below **The Pantanal is one of the largest floodplains in the world, and large areas are often inundated for up to seven months of the year, from November until May. In exceptional years (five in the 1900s), all but the highest ground is flooded, with predictably devastating results for the region's wildlife. As animals crowd onto the remaining land, vegetation is overgrazed and disease spreads rapidly.**

Right **The jaguar is the largest and most powerful of the American cats, and needs no special adaptation to take advantage of a wetlands habitat. Terrestrial mammals are the jaguar's main prey, although they frequently take caiman from the water.**

populations have declined dramatically in most areas, although in some of the more inaccessible areas caiman are still common. They have survived because, as well as being extremely secretive and wary, they have the ability to reproduce at a young age and adapt to different habitat types. These qualities will also help in restoring numbers if poaching can be controlled. Plans for developing a management programme for the species are already under way. Two government agencies, the Instituto Brasileiro de Meio Ambiente e dos Recursos Naturais Renovaveis (IBAMA) and Centro de Pesquisa Agropecuaria do Pantanal (EMBRAPA), are conducting investigations on the ecology of wild populations. Three private organizations are also involved in caiman studies.

Increased pressures

In conjunction with these efforts to stop wide-scale poaching, other actions are needed to address the wider problems facing the Pantanal. Although large tracts of the wetland remain remote and have been modified only slightly by man, recent acceleration in development pressure has given rise to growing concern. Deforestation in the watershed has increased siltation and altered the flooding cycle, while expansion of agriculture and associated dams and canals for irrigation have both made inroads on the wetland and altered flood flows. Industrial development, pesticide run-off, and contamination by mercury residues from gold mining, combine to create a serious pollution problem. The gold mining is a particularly serious problem both because of the toxic effects of the mercury and because the extraction process results in large quantities of sediment being flushed into the rivers, compounding the effects of deforestation.

In recent years, increasing anxiety about this array of pressures has led to state, federal, and international action to strengthen present conservation measures. However, because of the vested interests of many of the wealthy landowners, and the frequently expressed doubts about the efficacy of Brazil's environmental policies, much more work needs to be done before the spectacular riches of the Pantanal can be secured for future generations.

The Western Hemisphere Shorebird Reserve Network

The Western Hemisphere Shorebird Reserve Network (WHSRN) (see map) is a collaboration of private and government organizations committed to protecting the shorebirds of the Americas and the wetlands upon which they depend. Launched in 1985 to address conservation issues arising from research by the Manomet Bird Observatory (MBO), Philadelphia Academy of Natural Sciences and the Canadian Wildlife Service, WHSRN recognizes critical shorebird habitats and promotes cooperative management and protection of these sites. The network uses birds as a symbol for uniting countries in a global effort to maintain the Earth's biodiversity.

Each spring and autumn, many shorebirds migrate thousands of kilometres between their nesting grounds in the Arctic and their wintering areas in South America. Along the way they stop to feed and rest at a few widely separated wetlands, where they flock together in hundreds of thousands, or even millions, at one spot. If any one site were lost whole populations, and even species, could become extinct. Already habitats throughout the region are suffering from a variety of pressures and populations of some shorebirds have already declined by as much as 60–80 per cent in the last 20 years.

Education

In an effort to check the loss of shorebirds and their habitats, WHSRN brings critical sites into an international voluntary network. WHSRN and its affiliates, led by MBO and the National Audubon Society, then provide management recommendations and foster protection of these sites. They also sponsor training and education programmes with Latin American agencies, provide technical support and grants, and promote international research and twinning of geographically distant reserves. The enhanced prestige, assistance and international recognition gained from reserve status give strong impetus to effective long-term management and conservation of these critical wetlands.

Today, 18 reserves from Tierra del Fuego to Alaska protect around 30 million birds during their migration. Some of these were originally national or state parks, but others had no previous protected status, including some privately owned lands.

Copper River Delta

Stillwater — Great Salt Lake

San Francisco Bay
The Grasslands — Mono Lake — Cheyenne Bottoms

Bay of Fundy
Delaware Bay
Barrier Islands

● **Hemispheric:**
host at least 500,000 shorebirds

○ **International:**
at least 100,000

● **Regional:**
at least 20,000

☐ Countries participating

Bigi Pan — Coppename
Wia-Wia

Maranhao

Paracas ●

Mar Chiquita — Lagao do Peixe

Tierra del Fuego

Hidrovia

The Paraná-Paraguay river system has, for hundreds of years, provided an important transport artery into the heart of the South American continent. Today, encouraged by MERCOSUR, the common market established between the countries of the river basin, there are plans to alter the river so that it is navigable between Caceres in Brazil and the Port of Nueva Palmira in Uruguay. It is thought that such an alteration would allow the cheap transport of produce from the interior. However, while there is widespread enthusiasm for the project, which could do much for the commercial development of the region, there is also concern over its possible impact upon the natural systems of the basin, and over the economic costs of such changes. The most serious threat is to the Pantanal, where there are plans to dredge the

Right **The Pantanal in full flood.** At present, the flooding is an annual event that serves to store water from summer rains, mitigating the worst effects of the dry season. Ambitious plans to improve the navigability of the Paraná-Paraguay river system threaten this seasonal equilibrium. Large-scale hydro-engineering may alter severely the pattern of drainage so that some areas of savanna will become permanent swamp, and in other places grassland will be degraded. Although patches of wetland will still exist in something like their original state, there is growing concern that the Pantanal as we know it today will be destroyed forever.

Below **A mixed flock of black-necked swans** (*Cygnus melanocoriphus*) and coscorba swans (*Coscorba coscorba*) on the appropriately named Laguna de los Cisnes (Lake of Swans) in southern Chile. Other native waterfowl that breed in these temperate wetlands include the upland goose (*Chloephaga picta*), the ashy-headed goose (*C. poliocephala*) and dabbling ducks (*Anas* sp.).

River Paraguay for 670 kilometres (400 miles) between Corumba and Nuevo Caceres. This would not only result in major changes in the seasonal flooding cycle of the Pantanal and affect the region's rich animal and plant life, but it would also alter the pattern of flooding downstream. The Pantanal currently absorbs flood waters and releases these slowly during the course of the dry season. As a result, there is a difference of about four months between the floods of the Paraguay and Paraná rivers. If the Paraguay is dredged where it passes through the Pantanal, the flow of the river will accelerate and bring the timing of the flood peaks of the two rivers closer together. The effect would be an increase in the overall height of flooding downstream.

With these possible impacts there is currently growing international concern, and the funding agencies involved are being encouraged to carry out a detailed and independent analysis before the plan is approved and it becomes too late to reverse the decision.

Northern Europe

As the ice sheets moved back and forth over northern Europe, they created a heavily modified landscape with eroded hollows, glacial deposits and confused drainage. Today, the eskers (ridges of sand and gravel), drumlins (elongated, rounded mounds), terraces and deltas that are characteristic of Norway, Sweden and Finland are reminders of that glacial past. This post-glacial landscape provides the canvas for the most varied wetland regions in Europe.

Peatlands

Bogs and fens cover 60,000 square kilometres (23,000 square miles) in Sweden and Norway, often forming extensive mosaics together with wet forests. In Finland, the original area of bogs was about 110,000 square kilometres (42,500 square miles) – about 30 per cent of the land surface. Today, they are thought to cover less than 65,000 square kilometres (25,000 square miles). Their loss is due to natural drying of the peat substrate, drainage for agriculture and mining by the peat industry.

In general, bogs are more common in southern regions, while the largest fens are found in the northern and alpine regions. In summer, these wetlands form important breeding habitats for many northern shorebirds, swans (*Cygnus cygnus*) and cranes (*Grus grus*).

For centuries, peatlands have been an important resource in the rural economy. The reindeer herds of the Lapps graze on them in the summer, while substantial harvests of wild fruits are yielded, and some, like cloudberry, are harvested on a commercial basis. More recently, interest in the energy potential of peat has increased and in Sweden about 400 sites have been accepted for commercial exploitation. In Finland, 4 million tonnes of peat fuel are produced each year, as well as 300,000 tonnes of horticultural peat. This destructive use of peat adds to the pressure from agriculture and forestry which has resulted in the drainage of extensive areas over the centuries.

Iceland

The wetlands of Iceland support great numbers of breeding and migratory waterfowl, providing an important staging post in the flyways between the Canadian Arctic and Greenland and western Europe. The wide variety of important wetland habitats includes coastal mud flats, which are particularly important for geese, and inland marshes and lakes, which are essential for breeding ducks and waders. Over 4,600 sq km (1,800 sq miles) are considered to be of international importance and meriting special conservation status.

Sweden

In Sweden, acidification also poses a risk to human health, with a clear link being demonstrated between low pH values and high mercury concentration in fish, especially pike (*Essox lucius*). In more than 10,000 Swedish lakes, fish have passed the threshold of one part/million of their muscle containing mercury, and are unfit for human consumption. If acidification increases beyond present levels there is concern for those communities which depend upon groundwater for drinking supplies.

- Ramsar Sites
- Parks and Reserves
- Water Bodies
- Coastal Wetlands
- Deltas
- Peatlands
- Freshwater Marsh, Floodplains

Norwegian Sea

Myvatn-Laxá Region

ICELAND

Thjórsárver

Reykjavik •

ATLANTIC OCEAN

0 km · 250 · 500
0 mile · 125 · 250

Map Note

When compiling the West and Central European map, the lack of data has made it impossible to ensure that the peatland areas of Scandinavia are shown. A similar lack of data for the remaining floodplains of Europe has also resulted in incomplete coverage for these particular wetland systems.

Wet forests

Wet forests, dominated by birch trees (*Betula* spp.), form a transition zone between forest and open wetland habitats. They are key habitats for a large number of threatened or vulnerable animal and plant species, including many insects, mosses and lichens. Birds such as whimbrels (*Numenius phaeopus*) and great snipes (*Gallinago media*) breed there, and they provide a stronghold for mammals such as the European beaver (*Castor fiber*). In Sweden, wet forests and forested bogs cover approximately 50,000 square kilometres (19,000 square miles), and provide a habitat for almost 200 threatened species of flora and fauna. The growing recognition of wet forests as important for other values beside timber production has led to an intensive debate about the impact of modern forestry on natural habitats. As a result, subsidies for drainage operations have largely been withdrawn in Scandinavia, and a stricter legislation on ditching has been imposed on all land users.

Rivers and lakes

The Scandinavian peninsula has more than 100,000 lakes, in all covering more than 50,000 square kilometres (19,000 square miles); while in Finland, 10 per

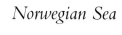

Orlandet · Tautra and Sv

8

NORWAY

Åkersvika

Oslo • Nordre Oye

Ilene and Presterodkilen

Kurefjorde

Ora

Jaeren

19

20

Skagerrak

Lake Hornborga

21 · 22

DENMARK

North Sea · Tipperne

30

Norway

In Norway, acidification is regarded as being responsible for the elimination of fish populations in waters extending over an area in excess of 13,000 sq km (5,000 sq miles). Birds and mammals are the most seriously affected as their available food supply is reduced. Some of these, such as the osprey (*Pandion haliaetus*), have important breeding populations in the region.

cent of the total area is covered by freshwater, including 60,000 lakes. Many of the Scandinavian lakes are deep, like Lake Vänern, but the majority are shallow and contain extensive areas of submerged and emergent vegetation. The lakes and watercourses provide critical wildlife habitats and an important recreational resource. Fishing and hunting are especially important. Many lakes also provide drinking water to urban areas and for industrial and agricultural purposes. For centuries, rivers have been important traffic arteries and used for logfloating in forest regions.

In the last 150 years, the use of these water resources has become less and less sustainable. A very high proportion of Scandinavian waters has been exploited for agricultural purposes and for power generation. In Sweden, for example, about 75 per cent of all suitable lakes and rivers have been regulated as part of hydroelectric developments, bringing irreversible ecological change. A large number of lakes have become totally or partly drained and turned into monotonous reedbeds or shrubby areas, thereby losing their original values. Today, the greatest problems in many areas are due to acid rain. Borne largely by the prevailing southwesterly winds, rain rich in oxides of sulphur and nitrogen falls upon the region. In Norway, about 40 per cent of lakes studied were reported to have a pH value of less than 5. In Sweden, more than 17,000 out of a total of 85,000 lakes have become significantly or strongly acidified, and the same general picture is true in Finland. As pH values fall below 5, many plant and animal species can no longer survive.

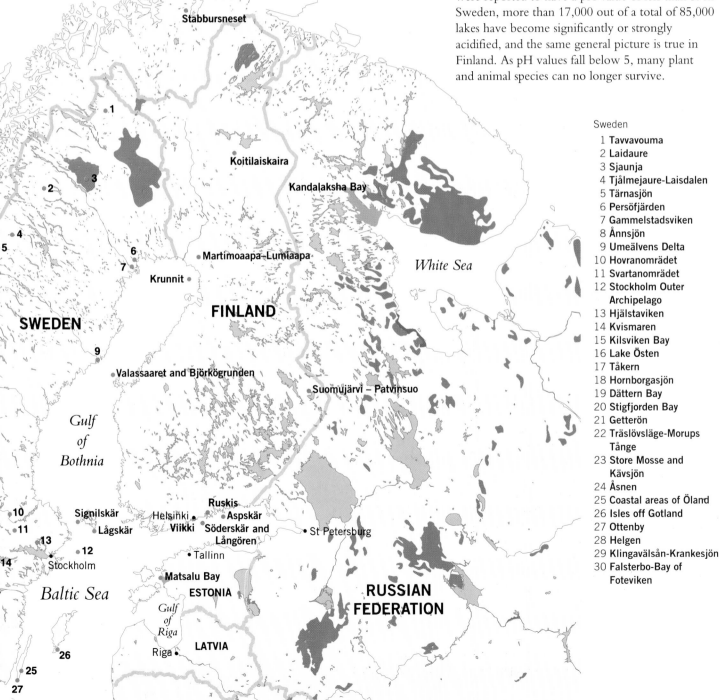

Sweden

1 Tavvavouma
2 Laidaure
3 Sjaunja
4 Tjålmejaure-Laisdalen
5 Tärnasjön
6 Persöfjärden
7 Gammelstadsviken
8 Ånnsjön
9 Umeälvens Delta
10 Hovranområdet
11 Svartanområdet
12 Stockholm Outer Archipelago
13 Hjälstaviken
14 Kvismaren
15 Kilsviken Bay
16 Lake Östen
17 Tåkern
18 Hornborgasjön
19 Dättern Bay
20 Stigfjorden Bay
21 Getterön
22 Träslövsläge-Morups Tånge
23 Store Mosse and Kävsjön
24 Åsnen
25 Coastal areas of Öland
26 Isles off Gotland
27 Ottenby
28 Helgen
29 Klingavälsån-Krankesjön
30 Falsterbo-Bay of Foteviken

Lake Hornborga

Over much of the 1800s and 1900s, Scandinavia and Finland suffered significant loss of wetlands. A programme that involved the lowering of lake water levels to increase the area of arable land was a major cause of loss, with more than a third of lakes affected in some regions of Sweden. One of the most famous of these lakes is Hornborga. By the mid-1900s, Lake Hornborga had been affected by five separate drainage operations. As a result, water level dropped by more than 2 metres (7 feet), and flooding was reduced to a seasonal phenomenon. Large parts of the former lake bed were colonized by bushes, others by extensive reedbeds, while the variety and number of waterfowl using the lake gradually decreased.

The problems were not, however, confined to wildlife. As a result of oxidation and desiccation of the organic soils, groundlevels fell and the costs of maintenance of the drainage scheme rose. Farmers began to ask for a reconsideration of the scheme and joined with conservationists who sought to restore the lake for its value as a bird habitat. After a lengthy process of scientific analysis and public debate, the restoration plan was accepted by the Swedish Parliament in 1989 and is expected to be complete by 1995. This will increase the wetland area from 27 square kilometres (10 square miles) to 34 square kilometres (13 square miles) and will cost a total of US$10 million.

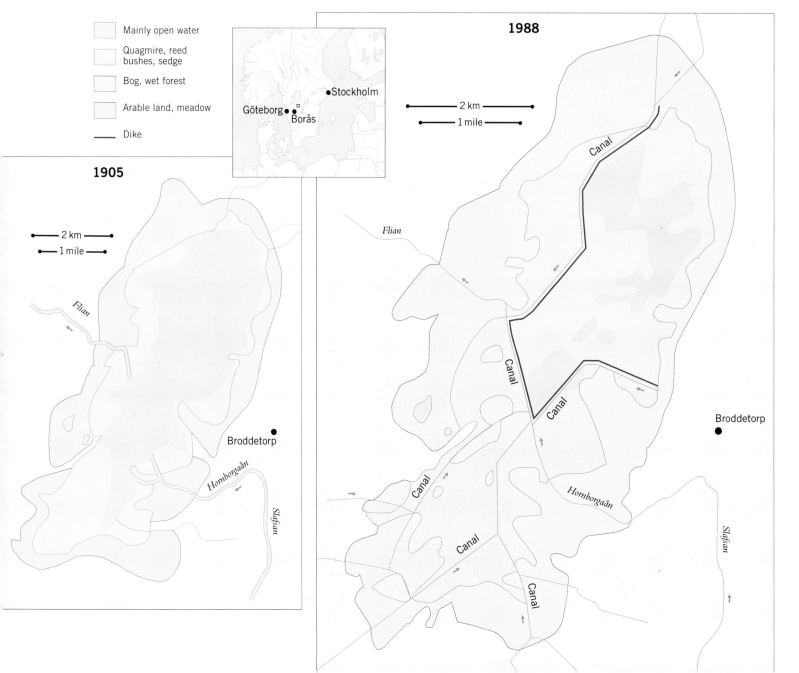

Denmark

Characterized by its low and gently undulating topography formed by glaciers during the last ice age, Denmark has a very large number of wetlands. Most lie in depressions along rivers, on the fringes of inland freshwater lakes and in coastal areas. Although most of them are small, their combined area and variety support a rich species diversity. In addition, on the edge of the surrounding shallow seas, the coastal fringe supports extensive shallow lagoons and fjords as well as salt marshes.

As is the case in the rest of northern Europe, the main threats to Denmark's wetlands are drainage and cultivation. Over the last 150 years these practices have reduced the country's freshwater wetlands by 75 per cent. Meanwhile, the land of the shallow coastal fringe is an easy target for land reclamation, and large coastal wetlands and shallow sea areas have been poldered and cultivated. However, today the need to reduce agricultural output has led to the adoption of set-aside policies which provide subsidies to farmers who maintain natural habitats. At the same time, growing awareness of nature conservation issues have helped to initiate several wetland restoration projects.

Tipperne

At the southern end of Ringkøbing Fjord, a large brackish lagoon of 300 square kilometres (115 square miles) on the west coast of Jutland, lies Tipperne, one of Denmark's oldest and most important wetland reserves. Covering 100 square kilometres (39 square miles), it is one of Denmark's 27 Ramsar Sites and has been designated as an EEC special bird protection area. Tipperne was established to protect breeding birds in the wet meadows, and create a refuge for migrating waterfowl.

However, Tipperne was not always such a haven for birds. The region's brackish, wet meadows are a semi-cultivated landscape, created and maintained by farming practices such as hay-making and grazing. Such practices, common up until the 1950s, were productive without fertilizers or draining programmes and they favoured the birds, with hay-making after the birds' breeding period and grazing by cattle and horses during autumn. At that time Tipperne was famous for its large breeding populations of birds such as ruffs (*Philomachus pugnax*), black-tailed godwits (*Limosa limosa*), avocets (*Recurvirostra avosetta*), redshanks (*Tringa totanus*) and gull-billed terns (*Gelochelidon nilotica*). Around 1950, however, a shift in the farming practices in the neighbourhood of Tipperne put an end to the need for hay and grazing areas, with the result that the vegetation grew high and dense, and the breeding populations of the birds dependent on the open wet meadow habitat decreased considerably. In response to this change in the ecological character of the site, subsidies were provided in order to reintroduce farming to the area. Today, during the hunting season, as many as 50,000 ducks seek protection within the confines of the reserve, while 20,000 ducks, 3,000–5,000 geese, and as many as 20,000 shorebirds use the wetland as a feeding area. In spring and summer, large densities of wading birds such as redshanks, black-tailed godwits and avocets have once again returned to Tipperne's meadows.

Left **The water level of Lake Hornborga in southwest Sweden was lowered five times in an attempt to gain more arable land. The map dated 1905 shows the lake just before the fourth lowering of the water level. The lake still comprised a large area of open water – surrounded by rushes and reeds, but still rich in bird life. By 1965, virtually no open water existed, and only sedges and reeds flourished. The birdlife declined dramatically. By 1988, restoration schemes had created small areas of open water, and birds returned to the lake.**

Above **Whooper swans (*Cygnus cygnus*) performing a so-called "triumph ceremony" at Pudassärvi in Finland. The display is usually performed by a male and female pair after aggressive encounters; it is also used as a greeting. The Finnish population of whooper swans suffered a dramatic decline in the 1900s, most likely due to illegal hunting and extensive plundering of eggs. By the 1950s the species was almost extinct. However, protective legislation has ensured their survival, and whooper swans have now regained their original range.**

Left **In recent years, large-scale clearing of vegetation has taken place in the northern parts of Lake Hornborga. This has caused the re-emergence of bodies of open water and increased the lake's attractiveness to birds.**

West and Central Europe

Stretching from the British Isles and the Brittany Peninsula to the plains of eastern Europe and the shores of the Black and Caspian seas, the middle latitudes of Europe contain some of the most densely industrialized and intensively farmed landscapes in the world. In the wake of the agricultural and industrial revolutions, and the centuries that followed, the region has lost the vast majority of its natural wetlands. In western Europe only small areas remain, on the lowland plains or in the high Alps, where montane wetlands are relatively undisturbed. The major exceptions to this, however, are the many estuaries and coastal flats, the floodplains of some of the major rivers, and the wetland mosaic of the Central European Plain. Although it has undergone substantial alterations in many areas, the plain has retained some of the most productive and biologically important ecosystems on the continent.

Coastal wetlands

Much of the European coast consists of a chain of extensive estuaries and intertidal bays separated by long stretches of rocky shore and sandy beaches. The Wadden Sea, which straddles the borders of Denmark, Germany and the Netherlands, is the largest, but areas such as the Wash, Morecambe Bay and the Solway Firth in Britain, and the Baie de Mont St Michel and the Baie de L'Aiguillon in France, are of major international importance. These wetlands are patchworks of sand and mud flats which support large populations of migratory shorebirds. Less visible, but equally important, is the role they play as nurseries for plaice (*Pleuronectes platessa*), sole (*Solea solea*), herring (*Clupea harengus*) and other species of commercially important fish, and their substantial harvest of mussels, cockles and other shellfish. Until now, the large size of many of these areas has prevented major damage. However, the Wadden Sea currently faces a myriad problems, while in Britain, the Severn Estuary would be altered substantially if a proposed power barrage was to go ahead. Several of the smaller estuaries, such as the Tees in England, and fringing salt marsh habitats have been substantially reduced in area as a result of wide-scale industrial development and agricultural encroachment.

All of Europe's major rivers once had extensive floodplains, but today only the Loire, Vistula, and parts of the Danube and Rhine retain a semblance of their former splendour. It is, therefore, not surprising that a 1987 symposium on "Conservation and Development of European Floodplains" concluded that floodplain ecosystems are among the most endangered in Europe. Dams and dikes to regulate flow and reduce floods have been the major cause of this wetland loss. However, with the growing awareness of the plight these ecosystems face, and a rising debate as to the economic, social and full environmental costs of their loss, many conservation groups in Europe are campaigning actively against the construction of further dams.

Conservation action

The World Wide Fund for Nature (WWF) has been particularly active and played a leading role in encouraging re-examination and cancellation of dams planned for the Danube and Loire, and in encouraging increased conservation investment. In 1991, WWF-Austria raised about US$8.5 million to purchase the 446 hectares (1,100 acres) of Regelsbrunner Au, part of the floodplain of the Danube downstream of Vienna. This secured the first component of the International Danube Park that will ultimately protect about 120 square kilometres (46 square miles) of floodplain wetlands from Vienna to Gyor in Hungary. Over 70,000 Austrians so far have purchased shares in the Regelsbrunner Au Reserve – a sizeable number of voters who would be antagonized if the Austrian Government were ever to attempt to appropriate the land for a reservoir.

The Central European Plain supports a diversity of wetlands, including peat bogs, fish ponds, freshwater marshes, wet meadows, soda lakes as well as the

Austria

1 Neusiedlersee
2 Stauseen am Unteren Inn
3 Donau – March – Auen
4 Untere Lobau
5 Rheindelta, Bodensee
6 Sablatnigmoor bei Eberndorf

Belgium

1 Vlaamse Banken
2 Zwin
3 Schorren van de Schelde
4 Kalmthoutse Heide
5 Ijzerbroeken
6 Marais d'Harchies

Denmark

1 Anholt Island Sea Area
2 Læso
3 Nordre Ronner
4 Hirsholmene
5 Randers and Mariager Fjords
6 Ulvedybet and Nibe Bredning
7 Vejlerne and Logstor Bredning
8 Nissum Bredning with Harboore and Agger peninsulas
9 Nissum Fjord
10 Stadil and Veststadil Fjords
11 Ringkøbing Fjord
12 Fiilso
13 Horsens Fjord and Endelave
14 Stavns Fjord and adjacent waters
15 Sejero Bugt
16 Næreå Coast and Æbelo area
17 Lillælt
18 South Funen Archipelago
19 Waters off Skælskor Nor and Glæno
20 Karrebæk, Dybso and Avno Fjords
21 Præsto Fjord, Jungshoved Nor, Ulfshale and Nyord
22 Waters southeast of Fejo and Femo Isles
23 Nakskov Fjord and Inner Fjord
24 Maribo Lakes
25 Waters between Lolland and Falster, including Rodsand, Guldborgsund and Boto Nor
26 Ertholmene Islands east of Bornholm
27 Vadehavet

Germany

1 Ostfriesisches Wattenmeer and Dollart
2 Hamburgisches Wattenmeer
3 Wattenmeer, Elbe – Wesser – Dreieck
4 Wattenmeer, Jadebusen and westliche Wesermündung
5 Lower Elbe
6 Muhlenberger Loch
7 Elbaue, Schnackenburg – Lauenburg
8 Krakower Obersee
9 Müritz See
10 Rügen/Hiddensee
11 Galenbecker See
12 Oder Valley, Schwedt
13 Peitz Ponds
14 Lower Havel and Gulper See
15 Steinhuder Meer
16 Weserstaustufe Schlüsselburg
17 Diepholzer Lowland Marsh and Peat Bogs
18 Dümmersee
19 Rieselfelder Münster
20 Unterer Niederrhein
21 Berga-Kelbra Storage Lake
22 Rhine, Eltville – Bingen
23 Lech-Donau Winkel
24 Water meadows and Peat Bogs of Donau
25 Ismaninger Reservoir and Fish-ponds
26 Lower Inn, Haiming – Neuhaus
27 Bodensee
28 Ammersee
29 Starnberger See
30 Chiemsee

IRELAND
Edinburgh
Solway Firth
Morecambe Bay
Lake District National Park (Esthwaite Water)
Shannon
Dublin
Severn
Thames
England

Marais du Cotentin et du Bessin (Baie des Veys)
Mont St Michel Bay
Brittany Peninsula
Golfe de Morbihan
Aiguillon Bay

Ramsar Sites
Parks and Reserves
Water Bodies
Coastal Wetlands
Deltas
Soda Lakes
Peatlands
Freshwater Marsh, Floodplains

great river floodplains of the Danube, Vistula, Dniepr, Dniester and Volga. Many of these valuable wetlands are still unprotected, although one of the most important sites, the Danube Delta, is now a Biosphere Reserve and listed under both the Ramsar and World Heritage Conventions. On the northern edge of the plain, the Baltic coast supports extensive areas of shallow water and coastal lagoons, which support large numbers of breeding, moulting, wintering and migratory birds.

Hungary
1 Lake Fertö
2 Tata, Old Lake
3 Kisbalaton
4 Lake Balaton
5 Velence – Dinnyés
6 Ocsa
7 Kiskunság
8 Hortobágy
9 Bodrogzug
10 Szaporca
11 Pusztaszer
12 Mártély
13 Kardoskút

Ireland
1 Lough Barra Bog
2 Meenachullion Bog
3 Pettigo Plateau
4 Easkey Bog
5 Owenboy
6 Knockmoyle/Sheskin
7 The Owenduff Catchment
8 Coole/Garryland
9 Mongan Bog
10 Clara Bog
11 Raheenmore Bog
12 Rogerstown Estuary
13 Baldoyle Estuary
14 North Bull Island
15 Pollardstown Fen
16 Slieve Bloom Mountains
17 Wexford
18 The Raven Nature Reserve
19 The Gearagh
20 Castlemaine Harbour
21 Tralee Bay

Italy
1 Palude Brabbia
2 Lago di Mezzola – Pian di Spagna
3 Lago di Tovel
4 Vincheto di Cellarda
5 Marano Lagunare – Foci dello Stella
6 Valle Cavanata
7 Torbiere d'Iseo
8 Valli del Mincio
9 Palude di Ostiglia
10 Isola Boscone
11 Laguna di Venezia: Valle Averto
12 Valle di Gorino
13 Valle Bertuzzi
14 Sacca di Bellocchio
15 Valle Campotto e Bassarone
16 Valle Santa
17 Punte Alberete
18 Piallassa della Baiona
19 Valli residue del Comprensorio di Comacchio
20 Ortazzo and adjacent territories
21 Saline di Cervia

Netherlands
1 De Boschplaat
2 De Griend
3 Zwanenwater
4 De Weerribben
5 Oostvaardersplassen
6 Engbertsdijksvenen
7 Het Naardermeer
8 De Biesbosch (part)
9 Oosterschelde
10 De Groote Peel

Slovakia
1 Súr
2 Cicovské mrtve rameno
3 Parízské mociare
4 Senné – rybníky

Switzerland
1 Les Grangettes
2 Baie de Fanel and le Chablais
3 Lac artificiel de Klingnau
4 Kaltbrunner Riet
5 Rade de Genève et Rhône en aval de Genève
6 Lac artificiel de Niederried
7 Rive sud du Lac de Neuchâtel
8 Bolle di Magadino

United Kingdom
1 Loch Eye
2 Lochs Druidibeg, a'Machair and Stilligary
3 Loch-an-Duin
4 Claish Moss
5 Rannoch Moor
6 Cairngorm Lochs
7 Loch of Skene
8 Loch of Lintrathen
9 Loch Leven
10 Loch Lomond
11 Feur Lochain
12 Glac-na-Criche
13 Bridgend Flats
14 Eilean Na Muice Duibhe (Duich Moss)
15 Silver Flowe
16 Gladhouse Reservoir
17 Fala Flow
18 Lindisfarne
19 Din Moss – Hoselaw Loch
20 Holburn Moss
21 Irtinghead Mires
22 Rockcliffe Marshes
23 Lough Neagh and Lough Beg
24 Esthwaite Water
25 Leighton Moss
26 Martin Mere
27 Derwent Ings
28 Rostherne Mere
29 The Dee Estuary
30 The Alt Estuary
31 Llyn Idwal
32 Llyn Tegid
33 Cors Fochno and Dyfi
34 Rutland Water
35 Ouse Washes
36 The Wash
37 North Norfolk Coast
38 Hickling Broad and Horsey Mere
39 Bure Marshes
40 Redgrave & S. Lopham Fens
41 Minsmere – Walberswick
42 Walmore Common
43 Upper Severn Estuary
44 Burry Inlet
45 Bridgwater Bay
46 Abberton Reservoir
47 Old Hall Marshes
48 The Swale
49 Chippenham Fen
50 Pagham Harbour
51 Chichester and Langstone Harbours
52 Chesil Beach and the Fleet
53 Exe Estuary

The Rhine

The Rhine River drains an area of 185,000 square kilometres (71,000 square miles) and flows for a total of 1,320 kilometres (790 miles) between the Alps and the North Sea. With a population of well above 50 million inhabitants, the Rhine Basin is one of the most densely populated regions in Europe, and is home to several heavily industrialized urban regions such as the Ruhr in Germany. The coal mines, steel mills and oil refineries of these centres lie at the heart of the European economy, and furnish industries producing machinery, electrochemical equipment, automobiles, construction materials, textiles, foodstuffs and paper. Huge chemical factories are also situated in the vicinity of Basel in Switzerland, Ludwigshafen, Frankfurt and Cologne in Germany, and Rotterdam in the Netherlands. In addition, a chain of conventional and nuclear power plants lines the river and its major tributaries.

Canalization

From the early 1800s onwards, the development of this economic heartland along the Rhine led to the loss and degradation of many of the wetlands and aquatic ecosystems of the basin. Most were lost because of huge investment in flood control and improved river navigation. The Upper Rhine floodplain between Basel and Karlsruhe was reshaped from a unique labyrinth of meandering river channels and wooded islands into a dull, monotonous landscape, with straight shipping canals, dikes and drainage ditches. With these drastic changes, the floodplain lost its natural value entirely and its suitability as a habitat for numerous animal and plant communities. Today, dams, weirs and other constructions have made the Upper Rhine and most tributaries inaccessible to migratory fish species such as salmon, which abounded a century ago.

In addition to the loss of biological diversity, these extensive modifications to the Rhine also reduced the floodplain's role in floodwater retention, thus increasing the risk of flood. Some 60 per cent of the floodplain, 15 square kilometres

Bottom left **The coal-fired power station at Herne in northern Germany. The picture shows how this particular stretch of the Rhine has been canalized to improve the transportation of coal. Although it was canalization such as this that cut journey times along the Rhine in half, the cost of improving navigation has been phenomenal. The former floodplains of the Rhine supported countless numbers of animal and plant species, as well as providing flood protection and water purification to thousands of people.**

Right **The canalization of the Rhine has been an epic civil engineering task spanning several generations. The three diagrams show how, in a period lasting just over 100 years, this section of the Rhine lost the vast majority of its floodplain behind dikes, dams and other constructions.**

Below **A desolate winter scene of pollution in Essen. Pollution reached a peak during the late 1960s and early 1970s, but following strict legislation, concentrations of pollutants have started to decrease.**

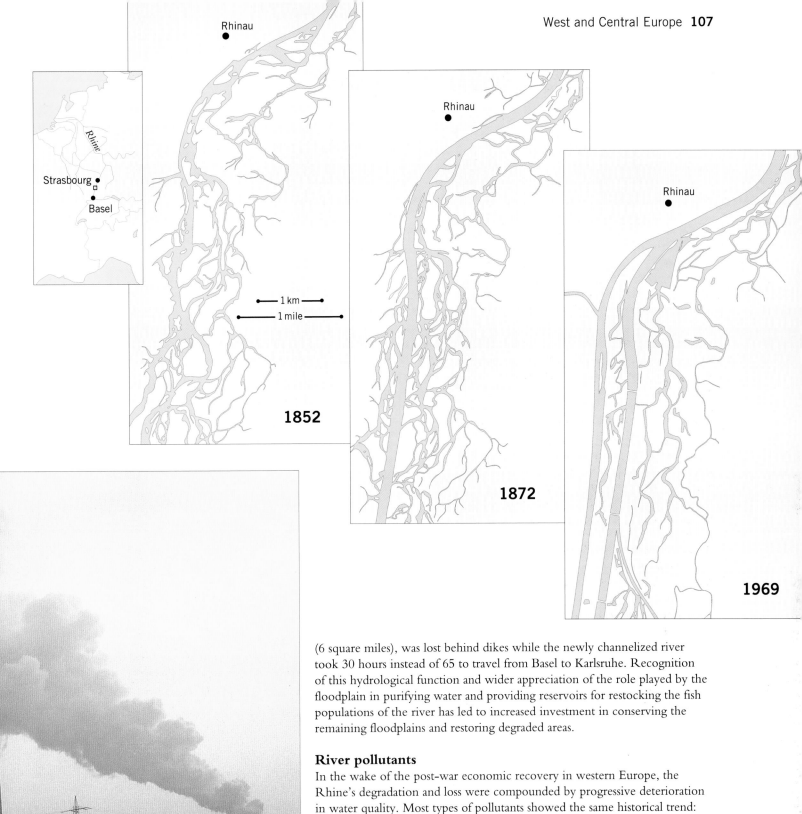

Rhinau

Strasbourg
Basel

Rhine

— 1 km —
— 1 mile —

1852

Rhinau

1872

Rhinau

1969

(6 square miles), was lost behind dikes while the newly channelized river took 30 hours instead of 65 to travel from Basel to Karlsruhe. Recognition of this hydrological function and wider appreciation of the role played by the floodplain in purifying water and providing reservoirs for restocking the fish populations of the river has led to increased investment in conserving the remaining floodplains and restoring degraded areas.

River pollutants

In the wake of the post-war economic recovery in western Europe, the Rhine's degradation and loss were compounded by progressive deterioration in water quality. Most types of pollutants showed the same historical trend: a rapid increase in the 1950s and 1960s, peak pollution between 1965 and 1975, and gradual recovery between 1975 and 1992. The recovery process is exemplified by decreasing concentrations of organic carbon, ammonium, heavy metals and other pollutants. As a result of diminished inputs of organic substances and ammonium, the oxygen levels improved gradually – an important prerequisite for the recovery of the river fauna.

Today, there are many signs that species diversity, namely invertebrates, molluscs and fish, is approaching pre-war levels again. This ecological recovery has not been cheap however. In the last two decades, thousands of municipal and industrial sewage treatment plants (STPs) have been built or upgraded. In the German Rhine Basin alone, DM30 billion were spent on sewer and STP construction between 1977 and 1986. In addition, industries have invested large sums in treatment facilities and clean technologies, either voluntarily or under the pressure of new environmental laws.

The Wadden Sea

The Wadden Sea is a shallow coastal wetland stretching for more than 500 kilometres (300 miles), from Den Helder in the Netherlands to the Skallingen Peninsula in Denmark. Covering an area of some 8,000 square kilometres (3,000 square miles), the tidal flats, sandbanks, salt marshes and islands which make up the Wadden Sea constitute Europe's largest intertidal wetland, and are the focus of substantial conservation investment from Denmark, Germany and the Netherlands.

A wildlife haven

Although its size, diversity and trans-nationality are sufficient cause for international concern, it is the Wadden Sea's role in supporting internationally important populations of animals and plants that have made it one of the world's most famous wetlands. Up to 12 million individuals of over 50 species of waterbird come here in the course of a year, most of them shorebirds from northwest Europe, Scandinavia, Siberia, Iceland, Greenland and northeast Canada. Similarly, most of the 102 species of fish which have been recorded from the Wadden Sea are only found there during certain periods of the year, migrating offshore or along the coast, according to their seasonal and reproductive cycles. Several of these fish are of major commercial importance: 80 per cent of plaice, 50 per cent of sole, and in some years a large part of the North Sea herring, for example, reach maturity in the Wadden Sea.

The common, or harbour seal (*Phoca vitulina*), the grey seal (*Halichoerus grypus*) and the bottle-nose dolphin (*Tursiops truncatus*) are all found in the Wadden Sea. Before 1980, hunting and pollution had exerted considerable pressure upon the population of the common seal. However, numbers had increased to 10,000 when, in 1988, an epidemic reduced the population by 60 per cent. There are strong indicators that PCBs (polychlorinated biphenyls) cause seals to become sterile and also weaken their immune system, making them more susceptible to infections.

Development pressure

The Wadden Sea lies on the coast of one of the most industrialized areas of Europe, and is under considerable environmental pressure. Construction of port facilities and embankments have resulted in damage to, and loss of, important habitats, in particular salt marshes. Although only a modest remainder of the extensive salt and brackish marshes, peatlands and lakes, which covered the area some 2,000 years ago, the salt marshes of the Wadden Sea are still the largest contiguous area of salt marsh in Europe. However, in the 50 years up to 1987, 33 per cent of their area was lost to embankments.

Pollution from the rivers which nourish the area, from the North Sea, and from the atmosphere, are of major concern. Of the five rivers entering the Wadden Sea, only the Varde Ao in Denmark is considered a fully natural estuary. The Ems, Weser, Elbe and Elder are all heavily influenced by large-scale engineering, harbour activities and dredging. All four are major sources of both nutrients and contaminants, with the Elbe being by far the biggest source. In addition, some dredge material dumped offshore contaminates the region with heavy metals. About a third of total lead inputs comes from this source. The atmosphere accounts for some 10 per cent of nitrogen and 25 per cent cadmium inputs to the Wadden Sea. Some progress, however, is being made in addressing these problems. Levels of phosphorus inputs appear to be decreasing slowly, but there is no clear decrease in nitrogen inputs.

Right **The island of Schiermonnikoog lies off the Netherlands' coast and is one of the five inhabited islands (out of a total of eight) that make up the West Frisian Islands. The West Frisians form a low-lying barrier between the Wadden Sea and the North Sea. Schiermonnikoog itself is some 16 km (10 miles) long and 3 km (2 miles) wide, and supports a few scattered farmstead. The low-lying brackish marshes are important sites for waterbirds.**

Above **With the exception of elephant seals, grey seals are the largest species of seal. Recent evidence has shown that grey seals and common seals, also found in the Wadden Sea area, are under risk from polychlorinated biphenyls (PCBs), chemicals once used in manufacturing processes. PCBs, once in the food chain, attack the seals' immune system, breaking down their defence against infection.**

Right **An aerial view of the Wadden Sea, literally the "shallow sea", shows the vast expanses of sandbanks and tidal flats for which the area is renowned. The region has attracted greater conservation interest than just about any other in Europe. The Wadden Sea supports more than 100 species of fish for at least one stage of their lives, some of which, such as plaice and sole, have huge economic value.**

Industry and conservation

Mussels have been harvested from the Wadden Sea for hundreds of years. But today this activity is mainly carried out on culture lots. Fishermen currently harvest seed mussels from other parts of the Wadden Sea and then spread these on the culture lots. In the long term this may result in the loss of natural mussel banks.

The tourist industry is another important source of income and employment in the Wadden Sea. While this brings many benefits it also threatens the fragile environment and for this reason controls on tourism are being developed. Most of the Wadden Sea has been made a protected area, and several sites have been included in the Ramsar Convention's List of Wetlands of International Importance. More important still, in 1982, Denmark, Germany and the Netherlands signed the Joint Declaration on the Protection of the Wadden Sea, which was designed to coordinate conservation activities. To assist in this work, the Common Wadden Sea Secretariat was established in 1987. The three countries have agreed upon a number of common objectives to reduce the ecological impacts of human activities. These common objectives include the prohibition of embankments, the closure of considerable parts of the Wadden Sea to mussel fishing, the establishment of special protection zones covering the most sensitive areas where no recreational activities are allowed, and the reduction by 50 per cent or more of the input of nutrients and pollutants between 1985 and 1995. For some particularly toxic substances, such as cadmium, it has now been agreed to reduce inputs by 70 per cent.

La Brenne

The Brenne is a land of lakes. Lying in central France, the region boasts over 1,100 lakes, covering an area of some 1,400 square kilometres (540 square miles). The lakes are fringed by extensive reedbeds and form one of France's most important, yet little known, wetland systems. Each spring, a wide range of rare and endangered waterbirds, notably bittern (*Botaurus stellaris*), purple heron (*Ardea purpurea*), black-necked grebe (*Podiceps nigricollis*), black tern (*Chlidonias niger*) and bearded tit (*Panurus biarmicus*) nest in the Brenne; while 15 species of amphibian, as well as 60 of the 90 species of dragonfly found in France, also breed here.

In view of the large area of wetland that has been lost because of human activity in other parts of Europe, it is ironic that the lakes of the Brenne are artificial, created by the construction of small dikes to hold back surface water. The first lakes date to the 1100s, when they were built by monks for fish production, most likely carp and pike. Fish farming remains the principal economic use of the lakes, with harvesting taking place in autumn or winter.

Le Mont St Michel

The bay of Le Mont Saint Michel includes 40 square kilometres (15 square miles) of salt marsh, the largest area on the French coast. Since the 11th century, farmers have used the salt marsh to graze sheep, which have subsequently become famous for the quality of their meat. These "moutons de pré salé" continue to form an economic resource, and today about 8,000 sheep are grazed on about 24 square kilometres (9 square miles) of the marsh. They share this with migratory birds, which visit the bay in winter, and breeding populations in summer. Brent geese (*Branta bernicla*) and widgeon (*Anas penelope*) are the two main species of waterfowl that graze the salt marsh during the winter.

Bottom **The historic fortress-abbey of St Michel towers over the surrounding salt marsh. Situated on a rocky offshore islet, Le Mont St Michel is linked to the mainland by a 1.5-km (1-mile) causeway. Although these wetlands have a limited, though sustainable, value in human terms (sheep farming), they provide an important feeding ground for migratory birds.**

Below right **The primeval beauty of a sunset over reedbeds in the Lac de Grand Lieu. Located in a sparsely populated region of western France, these wetlands have remained almost completely undisturbed. The lake is also fortunate in having no particularly rare or exotic species of wildlife, and has, for this reason, escaped the high-profile attention of tourists.**

Right **A grey heron hunts for fish in the shallows of the Lac de Grand Lieu. The isolated and tranquil waters of the lake have attracted the world's largest breeding colony of grey herons. The birds build their nests in the most inaccessible region of the lake, on the floating islands of reeds that drift slowly across the stretches of open water. Leaving their young in these secure island nests, the parents stalk the lakeside marshes for fishes and frogs.**

Lac de Grand Lieu

In contrast to the international fame of the Camargue, France's second largest wetland, the Lac de Grand Lieu in western France is little known both at home and abroad. Lying to the south of Nantes, the lake stretches over 75 square kilometres (29 square miles) and is bordered by flooded forest of willows and extensive reedbeds. Offshore, an expanse of water lilies alternates with forested islands; true floating islands of trees, each one moving on its muddy base. Grand Lieu is a world apart, isolated from the surrounding countryside by its watery medium. In spring, a chorus of frogs drowns out human voices, while grey herons (*Ardea cinerea*) and little egrets (*Egretta garzetta*) move between the floating islands where they nest and the surrounding marshes where they feed.

Grand Lieu supports the world's largest colony of grey herons, of which there are 1,300 pairs in the region. In addition, in winter 17,000 ducks roost on the lake during the day, moving to feed during the night in the Baie de Bourgneuf. A small community of about 20 fishermen use the lake, using nets and traps to capture perch (*Percha fluviatis*), pike (*Esox lucius*), European eel (*Anguilla anguilla*), carp (*Cyprinus carpio*) and tench (*Tinca tinca*), all of which are valued highly locally. In 1980, a reserve of 27 square kilometres (10 square miles) was established, and efforts are under way to expand this further.

Sava River Valley and Kopacki Rit

Lying just south of Zagreb on the border of the republics of Croatia and Bosnia, the floodplain wetlands of the Sava River cover an extensive area. As well as being a biosphere reserve, the region also contains three bird sanctuaries. The Sava wetlands, once a mosaic of alluvial forests known as the "Slavonian Oak Forests", have been reduced by drainage and land reclamation at the periphery. Of the original alluvial forest wetlands, 600 square kilometres (230 square miles) remain, of which 70–80 per cent are still inundated. There are 200 square kilometres (77 square miles) of water meadows, which are cut once a year and provide important pasture land. Also important are 15 square kilometres (6 square miles) of fish ponds with shallow water, reedbeds and mud habitats. Permanent water lies in depressions and oxbows, for instance at Krapje Dol, Vrasje Plato and Rakita.

Lying between the Danube and Sava rivers is Kopacki Rit; 500 square kilometres (190 square miles) of floodplain, of which 72 square kilometres (28 square miles) are strictly protected and 105 square kilometres (40 square miles) as a nature park between the Drava and Dumar rivers in Croatia.

Wildlife and threats

The oak forests support significant numbers of black stork (*Ciconia nigra*), white-tailed eagle (*Haliaeetus albicilla*) and lesser-spotted eagle (*Aquila pomarina*), together with 250 pairs of white stork (*Ciconia ciconia*), 150 pairs of European spoonbills (*Platalea leucorodia*) and 450 of grey heron. Mammals include large numbers of red deer (*Cervus elaphus*), wild boar (*Sus scropha*) and otter (*Lutra lutra*). Comparable in ecological importance to the Camargue in France and Coto Doñana in Spain, the riverine forests are possibly the best that remain in the whole of Europe.

The future of the floodplain looks uncertain because only relatively small parts of the area are protected, and it is likely that any external interference will damage the area as a whole. Rearrangement of drainage patterns outside areas such as Mokro Polje and Lonjsko are examples of this type of interference. Drainage and water pollution are major concerns, as is the intensification of agriculture, while nearby industrial development has led to contamination by heavy metals.

The civil war that started in 1991 also has had a major impact on pollution. For example the Sisak petroleum refinery, lying 60 kilometres (40 miles) southeast of Zagreb, was repeatedly hit, releasing several hundred tonnes of oil, diesel fuel and aromatic compounds into the Sava River.

Below **An area of Kopacki Rit in Croatia during the seasonal spring flood. The channel created by boats cuts a swathe through the ubiquitous pond weed that surrounds the "Slavonian" oaks, the trees for which this region of the Danube floodplain is famous.**

Top right **Black storks were once more common throughout former Yugoslavia, but the 1900s witnessed a decline of the species, particularly in the north of the region. The floodplain oak forests of the Sava and Danube rivers, however, still support a significant number. This wetland habitat provides the black storks with their chiefly fish diet.**

Above **The Danube River at Kazan Gorge, southern Romania. Increased pollution of the Danube, particulary over the last 40 years or so, has prompted growing international concern and collaboration to develop a comprehensive programme of action to address this problem.**

Right **The wild boar will usually choose damp, forest regions as its preferred habitat. Here it can forage for a wide array of food, from leaves and fruits to frogs and mice, in the moist soil conditions.**

The Danube

Draining a basin of 805,300 square kilometres (310,800 square miles), and flowing 2,860 kilometres (1,716 miles) to its delta on the edge of the Black Sea, the Danube is the second largest river in Europe after the Volga. In its lower reaches in Romania, the Danube formerly flooded over 11,000 square kilometres (4,250 square miles) of wetlands, about 5 per cent of the country's land surface. These vast wetlands yielded a variety of benefits and played an especially important role in reducing the risk of flood, being able to retain up to 9 cubic kilometres (1.3 cubic miles) of water, and serve as a filter between the basin and the Black Sea.

At 5,200 square kilometres (2,000 square miles), the delta is one of the largest wetland systems in Europe. Characterized by its wilderness of reedbeds, seemingly endless maze of canals, lakes and ponds lined with white willows and poplars, and the fossil dunes with their mosaic of forests and sandy grasslands, the delta forms one of Europe's unique habitats. It supports a rich diversity of plant and animal species, many of which are seriously threatened or have been lost elsewhere in Europe. Bird populations are exceptionally rich, with some 280 species nesting, resting and feeding in the area.

Modern impacts

Like most large European rivers, the Danube has suffered from pollution, modification of the hydrological system and drainage of its wetlands. Over 4,000 square kilometres (1,500 square miles) – 69 per cent – of the floodplains of the lower Danube, and 800 square kilometres (300 square miles) – 15 per cent of the delta – have been drained, while dredging of canals within the delta have reduced its capacity to retain and filter water. At the same time industry, agriculture, livestock and urban settlements have all increased the input of sewage, agricultural and industrial waste and pesticides into the Danube's water. This has increased substantially the nutrient load of the river, by up to three times in the case of dissolved nitrogen, and up to 12 times in the case of phosphorus. This combination of wetland loss, flow alteration and pollution have together resulted in substantially increased eutrophication and a marked reduction in the productivity of the river system.

In response to these problems, the new Romanian Government began, in 1990, to work with UNESCO, IUCN, WWF and a number of other international organizations to conserve the biological wealth of the river and its delta, in particular by improving water quality in the basin and restoring some of the former wetland system. A regional project to improve water quality in the Danube Basin is also now under way, and a management plan for the delta is in preparation.

Drainage

In most parts of northwestern Europe, the area of semi-natural grazing marsh and wet meadow has declined sharply in the last 40 years, a trend which is probably more attributable to drainage and agricultural improvement than any other cause. In Belgium, for example, drainage of semi-natural wetlands has taken place at the rate of 10–12 square kilometres (4–5 square miles) a year since 1960. Flooded pastures in river valleys covered about 350 square kilometres (135 square miles) in Flanders in 1960 but has now been reduced to less than 10 per cent of this area. Whereas the scale of wetland reclamation is now much reduced and it is no longer perceived as a major threat, there are still individual sites endangered by drainage.

The general pattern in Belgium, with drainage rates falling but certain individual sites under threat, is typical of much of the United Kingdom,

Below **Mechanized drainage at Woodwalton Fen in Cambridgeshire, England. In the last 40 years, drainage has destroyed vast areas of marsh, fen and other natural wetlands all over western Europe. Drainage, undertaken to create agricultural land, can drastically lower the water table over a large area.**

Right **Common blue damselflies (*Enallagma cyathigerum*) are found in many European wetlands. When mating, the male (left), after transferring sperm to the lower part of the abdomen, holds the female's neck with the end of the abdomen. This causes the female to lift her abdomen and collect the sperm from the male.**

Ireland, the Netherlands and Germany. In England, landscape surveys suggest that the area of freshwater marsh fell by about 52 per cent between 1947 and 1982, with smaller losses in Wales. Drainage resulted in the annual loss of around 40–80 square kilometres (14–28 square miles) of damp grassland and marsh in England and Wales in the 1970s and early 1980s. Arterial drainage projects, which lower water tables over a substantial area, precipitated some of the most significant habitat losses. However, during the late 1980s and early 1990s there was a sharp fall in the rate of arterial and field drainage, partly because of a cut in the grants available for such work. And where schemes did go ahead, conservation measures took a higher priority than in the past. Northern Ireland is the main exception to this trend. Here drainage rates remained relatively high, at least until 1986, with many works being environmentally damaging.

In Germany, as in the Netherlands, most wetlands with potentially productive soils have been drained already, although small wetlands are still under pressure from agricultural improvement. National statistics suggest that the area of bogs and fens within farmland was still falling in the 1980s, being reduced from around 1,170 square kilometres (450 square miles) in 1981 to just over 1,070 square kilometres (410 square miles) in 1985. Several Ramsar Sites in Germany are under threat from agricultural or recreational pressures; for example parts of the Danube water-meadows are drying out because of intensive agriculture in the surrounding area.

Right **Turf Fen Mill lies by the River Ant in Norfolk, England. This tranquil scene belies the many threats that the Norfolk Broads face today. Ironically, it is scenes such as this that encourage thousands of tourists to visit the region each year. The sheer weight of numbers and the wide variety of activities they pursue is disturbing the region's fragile ecosystem.**

The Norfolk Broads

The Norfolk Broads are a collection of small lakes along the valleys of the lowland rivers which drain eastern Norfolk. Meaning broadening or widening of a river, the Broads are famous for their reed-fringed rivers and lakes, the gentle contours of the surrounding farmland and the many picturesque thatched cottages. Yet, only a few of the thousands of people who come to visit each year realize that this is a man-made ecosystem. The lakes all lie on peat deposits that were excavated for fuel during the 1300s and 1400s. The workings were abandoned when they were flooded due to a rise in sea level and a wetter climate. However, the surrounding marshes and fens continued to be used for hunting of wildfowl, fishing and harvesting reeds (*Phragmites australis*) and sedge (*Cladium mariscus*) for roofing material. This led to the progressive digging out of channels or dikes connecting the lake basins with the rivers so that the marsh products could be transported easily by boat to the surrounding villages and towns.

An invaluable asset

Today, the Broads represent many different things to many people. Local people see the wetlands as adding to their household economy; for some they provide fertile land on which to earn a living, while for others they are a source of water and a means of sewage disposal; in addition the region is a place of recreation for those who sail, fish, walk and holiday in its waterways and marshes, while for naturalists it harbours ecosystems which provide a varied plant and animal life.

Over the past 45 years, the Broads have undergone many environmental changes, most of which have been detrimental to the area. There has been widespread loss and degradation of natural habitats, most notably in the water-plant life of the rivers and broads. And the impoverishment of fenland plant communities as a result of neglect and alteration of water cycles has worsened the situation. Further damage has been caused by both the conversion of many of the traditional high-water table grazing marshes into deep-drained arable agriculture, and progressive erosion of the river banks and protective flood walls as a result of boat wash. Uncontrolled mooring and trampling by the hundreds of anglers that visit the Broads is also a problem.

In addition to their more obvious ecological effects, these changes have had important economic consequences. It is already proving costly to maintain the fragile ecosystems of the region as they are and, if the degradation is not halted, the cost of repairing the floodwalls, piling the eroded river banks and protecting key habitats could prove to be exorbitantly expensive.

In pursuit of this task, the Broads Authority is building consensus in support of a Broads Strategy and Management Plan. The plan is designed to restore the Broads to ecological health and economic wealth, and at the same time to carry public support and approval for the required policies and actions.

Mediterranean Basin

The empires of Ancient Greece and Rome ruled much of the
Mediterranean it was a region rich in wetlands. The wetlands remained
largely intact until the 1800s and 1900s, when most were drained either
to provide agricultural land or to eradicate malaria. Today, the remaining
wetlands of the Mediterranean comprise an estimated 6,000–8,500 square
kilometres (2,300–3,300 square miles) of coastal lagoons, between 8,000 and
10,000 square kilometres (3,000–3,850 square miles) of natural lakes and
marshes, mainly lying in river deltas and the region's remaining floodplains,
and over 10,000 square kilometres (3,850 square miles) of artificial wetlands,
mostly reservoirs.

The deltas of the north shore of the Mediterranean support some of the
most extensive and varied wetland areas remaining in the whole of the
Mediterranean Basin. The Ebro in Spain, Rhône in France, Po in Italy,
Axios and Evros in Greece, Menderes, Seyhan and Ceyhan in Turkey are
all complex mosaics of wetland and dryland habitats, including dunes with
juniper and pine trees, lagoons, freshwater and brackish marshes, and
freshwater lakes.

Along the southern Mediterranean shore of North Africa, most rivers
are short, highly seasonal and do not provide sufficient sediment for delta
formation, the Nile being the only exception. However, the once mighty
delta of Africa's longest river has been degraded by the Aswan Dam.

Vanishing wetlands

Coastal lagoons, such as those of the Languedoc in southern France, are the
most typical wetlands of the basin and are scattered in an irregular fashion
along its shores. Most are connected to the sea by narrow channels, which are
either artificial or natural, permanent or temporary, and vary in salinity from
season to season.

Riverine floodplains have been reduced to a few tiny isolated remnants
as the result of 2,000 years of hydraulic engineering. Oxbows and floodplain
freshwater marshes are now rare as most have long since been drained for
agriculture. The only surviving examples are found in a fragmentary and
highly modified state in the Languedoc and Crau regions in southern France,
in the Po Valley in northern Italy, in northern Algeria, and in the now much
degraded Parque Nacional de las Tablas de Daimiel on the River Guadiana in
central Spain. Along with the destruction of the floodplains, diking, grazing,
agriculture and felling for timber have killed the riverine forest. Only a few
isolated relict stands still exist in the Mediterranean region, such as the Nestos
Delta in Greece and around the Moraca River in Yugoslavia.

Natural freshwater lakes are also infrequent in the Mediterranean region
except for a few examples such as Lago Maggiore and Lago di Garda in Italy,
Lakes Mikri and Megali Prespa in Greece, and others in Yugoslavia, Albania
and Turkey. In North Africa, permanent freshwater lakes have always been
scarce outside of the sub-humid mountainous areas, and the largest examples
have now been drained, such as Lake Fetzara in Algeria. The only large
lowland freshwater lakes that remain either dry out periodically or are in
coastal areas and, such as Lake Ichkeul in Tunisia, receive some inflow of
sea water during the drier periods of the year and in times of drought.

Several countries have inland salt lakes. These vary from sites lying below
present sea level such as Chott Melrir in Algeria, to upland basins at over
1,000 metres (3,300 feet) such as the Plain of Chotts in Algeria, the Laguna
de Gallocanta in Spain and Lake Tuz in Turkey. All of these salt lakes are
shallow and dry out completely from time to time.

Conservation issues

Just about every type of wetland in the Mediterranean
region has been exploited by man since the beginning

Greece

1 Lakes Mikri Prespa and Megali Prespa
2 Axios – Loudias – Aliakmon Delta
3 Kerkini Reservoir
4 Lakes Volvi and Langada (Koronia)
5 Nestos Delta and Gumburnou Lagoon
6 Lake Vistonis and Porto Lagos Lagoons
7 Lake Mitrikou and adjoining Lagoons
8 Evros Delta
9 Amvrakikos Gulf
10 Messolonghi Lagoons
11 Kotichi Lagoon

Italy

1 Palude Brabbia
2 Lago di Mezzola – Pian di Spagna
3 Lago di Tovel
4 Vincheto di Cellarda
5 Marano Lagunare – Foci dello Stella
6 Valle Cavanata
7 Torbiere d'Iseo
8 Valli del Mincio
9 Palude di Ostiglia
10 Isola Boscone
11 Laguna di Venenzia: Valle Averto
12 Valle di Gorino
13 Valle Bertuzzi
14 Sacca di Bellocchio

0 km	250	500
0 mile	125	250

- Ramsar Sites
- Water Bodies
- Coastal Wetlands
- Seasonal Wetlands
- Deltas
- Soda Lakes
- Freshwater Marsh, Floodplains

15 Valle Campotto e Bassarone
16 Valle Santa
17 Punte Alberete
18 Piallassa della Baiona
19 Valli residue del
   Comprensorio di Comacchio
20 Ortazzo and adjacent
   territories
21 Saline di Cervia
22 Padule di Bolgheri
23 Diaccia Botrona
24 Laguna di Orbetello
25 Lago di Burano
26 Padule di Colfiorito
27 Lago di Nazzano
28 Lago di Villetta Barrea
29 Lago di Fogliano
30 Lago dei Monaci
31 Lago di Caprolace
32 Lago di Sabaudia
33 Saline di Margherita di Savoia
34 Torre Guaceto
35 Le Cesine
36 Bacino dell'Angitola
37 Stagno S'Ena Arrubia
38 Stagno di Mistras
39 Stagno di Sale Porcus
40 Stagno di Cabras
41 Corru S'Ittiri Fishery – Stagno

di San Giovanni e Marceddi
42 Stagno di Pauli Maiori
43 Stagno di Cagliari
44 Stagno di Molentargius

Spain
1 Ria de Ortigueira y Ladrido
2 Umia-Grove, La Lanzada,
   Punta Carreiron y Lagoa
   Bodeira
3 Lagunas de Villafáfila
4 Las Tablas de Daimiei
5 S'Albufera de Mallorca
6 Laguna de la Vega (o del
   Pueblo)
7 Prat de Cabanes – Torreblanca
8 L'Albufera de Valencia
9 Marismas del Odiel
10 Doñana
11 Lagunas de Cádiz (Medina y
   Salada del Puerto)
12 Lagunas del Sur de Córdoba
   (Zóñar, Amarga, Rincón)
13 Laguna de Fuentepiedra
14 Salinas del Cabo de Gata
15 Salinas de la Mata y
   Torrevieja
16 Pantano de el Hondo
17 Salinas de Santa Pola

of civilization. Among the most common human uses have been drainage and conversion to agriculture, grazing, water storage, fisheries and aquaculture, mineral exploitation, hunting, harvesting of wetland vegetation, tourism and water sports. In recent years, urban and industrial development have also increased substantially, adding greatly to the rate of wetland loss, much of it through drainage for building land. Because of economic development and a rapidly increasing population, the present demand for water often exceeds the available resources in the Mediterranean.

The conservation, protection and management of water resources is fast becoming a national priority in all the countries of the Mediterranean Basin. This inevitably causes conflicts of interest in the use of this resource and in the management of wetland areas. However, most countries have made an explicit commitment to wetland conservation by ratifying the Ramsar Convention. Only Albania, Turkey, Cyprus, Lebanon, Syria, Israel and Libya have so far failed to sign the Convention. Still, the future for Mediterranean wetlands and their flora and fauna is uncertain, and far greater investment in conservation will be needed if the increasing public and government interest is to halt the loss of the region's wetlands.

Flamingos of the Camargue

The Rhône Delta region, better known as the Camargue, is host to hundreds of thousands of visitors each year; some are on vacation, others on day trips to the beach. As they drive across this flat landscape most hope to see the black bulls, white horses and, above all, the "pink" or greater flamingos (*Phoenicopterus ruber roseus*), which, in the mind of most visitors, characterize this most famous of French wetlands. Flamingos have become a symbol of conservation success in France. Since records began in 1944, the number of flamingos breeding in the Camargue has grown from about 3,000 in 1947 to 20,000 in 1986. Since 1981, at least 8,000 pairs have nested each year. In winter too, the population has risen, with the number wintering in France growing from 500 in 1969 to 24,000 in 1991.

The reasons for this increase in both breeding and wintering populations are complex. However, a central factor has been the presence of a secure nesting and feeding site in the salines (small salt pans and basins) of the Camargue.

Although some flamingos feed within a 10-kilometre (6-mile) radius of the saline region, others fly to feeding sites up to 70 kilometres (40 miles) away along the coast of southern France, returning every two to four days to relieve their incubating partner, and in due course feed their single chick. Since 1974, breeding birds have used an artificial earthen island in the Etang de Fangassier, where they are protected from disturbance and from which over 80,000 chicks have been raised.

Elsewhere in the western Mediterranean, flamingos only breed regularly (about once every two years) in the lagoon of Fuente de Piedra, near Malaga in southern Spain. These birds fly up to 150 kilometres (90 miles) to feed in the Coto Doñana in the delta of the Quadalquivir River, south of Seville. Although flamingos have bred in Coto Doñana in the past, the absence of a secure nesting island has limited breeding to only five occasions since 1970.

The salines

Salt has for centuries been an important trading commodity and the basis of major trade routes throughout history. One of the easiest ways to obtain salt is to allow sea water to evaporate naturally using the heat of the Sun and then collect the residue. This technique is only really economically viable in areas with long periods of warm weather and little precipitation – the long Mediterranean summer is ideal and many salines produce salt for the table and for industry in this way.

Salines are composed of small basins and salt pans for the production of salt; in total, they cover some 1,000 square kilometres (400 square miles) and exist in virtually all countries of the Mediterranean region. Some are run as state or privately owned industries, others are smaller, traditional and less efficient. The largest saline covers 110 square kilometres (42 square miles) and lies in the south of the Camargue. All salines work in the same way. Sea water is moved, or pumped, through a series of pans, becoming increasingly saltier due to evaporation as it approaches the final settling pans where the salt crystallizes. This circuit might take anything up to two years to complete in a large saline.

As the water moves through the system, the regular flooding and drying of the shallow pan creates ideal conditions for certain algae and invertebrates, providing rich feeding grounds for birds. The brine shrimp *Artemia*, for example, is particularly important, and is a key component of the diet of many bird species, in particular flamingos. In addition to their high productivity, the presence of many ponds of different salinity at any one time allows many different species of bird to coexist, while the stability of the pumping cycle from year to year means that conditions are usually predictable. This stability and predictability are of special value in the Mediterranean, where wetland conditions tend to be extremely variable and the lack of a strong tidal cycle makes mud flats few and far between. For these reasons salines are key breeding and feedings sites for flamingos, avocets (*Recurvirostra avosetta*), little terns (*Sterna albifrons*), slenderbilled gulls (*Larus genei*), Kentish plovers (*Charadrius alexandrinus*), gull-billed terns (*Gelochelidon nilotica*) and sandwich terns (*Sterna sandvicensis*). Between 50 and 80 per cent of all flamingos in southern France feed in salines during the summer months.

Salines are a good example of how economic interests can coexist with, and even benefit, local wildlife. However, small salines are increasingly uneconomic and their closure may have considerable local consequences for bird populations. In Portugal, for example, two-thirds of the 300 salines are either already closed, or threatened with closure. Conservationists do not at present have the resources to keep salines functioning purely for their ornithological interest and importance.

Above left **Greater flamingos perform their characteristic mating display in the Camargue in France. The salt pans of the region, as well as providing a stable supply of algae and invertebrates on which flamingos feed with their specially adapted straining beaks, also provide a secure artificial nesting site. These near perfect conditions have given rise to dramatic increases in the numbers of birds in recent years.**

Above **Along with the flamingos, the white horses of the Camargue attract many thousands of visitors to the region each year. These horses are thought to be descended from Arab stock brought to the region by the Saracens.**

Left **The process of evaporation can clearly be seen in this picture of salines in the Camargue. The bottom right-hand saline contains the brackish seawater at an early stage of the process. When it reaches a certain salinity it is pumped into the neighbouring saline, in which the concentration of salt continues to increase. By the time the water has reached the top right-hand saline, salt ridges are clearly defined, and the process of salt production is almost complete.**

Traditional Fisheries and Aquaculture

Rivers, reservoirs, lakes and coastal lagoons contribute about 25 per cent of the total commercial fishing undertaken within the Mediterranean Basin. Of these four wetland types, lakes and coastal lagoons are the most important, with the largest number of fish caught in lagoons. Fishing in lagoons is generally seasonal, taking place in autumn and spring. The main target species include eels, grey mullet (*Mullus* sp.), sea bass (*Dicentrarchus labrax*) and sea bream (*Pagellus bogaraveo*), all of which spawn in the sea and use the relative safety of the lagoons for feeding and growth.

Over the past 20 years, these traditional fisheries have declined and aquaculture has become increasingly important. The main shellfish species farmed are mussels, oysters and clams, although, more recently, cage culture of sea bass and sea bream, and pond culture of prawns and shrimps has been developed.

Often fish farming has been recommended in order both to offset the decline in traditional fisheries and to make the maximum profit from wetlands. However, there is now a strong argument against these uses on both environmental and economic grounds. In many areas, aquaculture has increased natural wetland degradation by aggravating eutrophication and altering the habitat through the construction of dikes and sluices. In addition, fish-farming activities usually require large quantities of high-quality fresh water, adding to the already large number of users for this precious, overstretched resource.

The long-term economic viability of fish-farming is unclear. Initially, many farmers received massive financial aid from national and international development agencies, but most of the aid has now been withdrawn. Other farmers face serious economic difficulties because bacteriological contamination of their stock has prevented sales. Contamination has become increasingly frequent in recent years, especially in sites within lagoons.

Turkey's Wetland Diversity

Surrounded by water on three sides – the Black Sea to the north and the Mediterranean to the south and west – Turkey is home to some of the most important wetland resources in the Mediterranean. Two large deltas are situated on the Mediterranean coast, one on the Aegean, and another on the Black Sea. Most coastal wetlands consist of extensive lagoon systems that are not only important for wildfowl but also for the valuable local fisheries. Inland, Turkey has a variety of wetland types. There are, for example, many deep freshwater lakes – especially in the "lake district" of western Anatolia; the central plateau has extensive marshes and shallow lakes, while eastern Turkey is a region of soda lakes and river meadows. One of the country's most important lakes is Burdur Golu, which in winter supports up to 70 per cent of the world's population of white-headed duck (*Oxyura leucocephala*).

On the south coast of Turkey, the Goksu Delta is home to the majority of the Turkish population of purple gallinule (*Porphyrio porphyrio*) and a large breeding population of the marbled teal (*Marmarcetta angustirostris*). The delta is also an important stop-over for white storks (*Ciconia ciconia*) and white pelicans (*Pelecanus onocrotalus*), white-tailed eagles (*Haliaeetus albicilla*) and spotted eagles (*Aquila clanga*) as well as tens of thousands of wintering waterfowl. The delta became a protected area in 1990.

Left The fire salamander (*Salamandra salamandra*) is a largely terrestrial animal that returns to water to mate. However, it has to keep its skin moist, and often shelters in damp logs. If the logs are put onto a fire, the salamander will rush out. This has given rise to the folk-tale that salamanders live in flames and fire.

Left Local fishermen try their luck in one of the many coastal lagoons of the Mediterranean. Several species of marine food-fish use these lagoons as nurseries. After hatching out at sea, the fry migrate into the lagoons, which provide a secure nursery environment. Traditional fishing methods that harvested wild fish did little harm to the environment. The recent introduction of commercial fish-farming, however, presents a greater environmental threat.

Below Fishing boats moored on the shores of Lake Egridir in western Turkey's Anatolian "lake district". The presence of a fleet of fishing boats, albeit very small ones, shows that this inland lake still provides the local inhabitants with an exploitable harvest of freshwater fish. Although the numbers of fish caught has been declining steadily for some years, these inland fisheries provide both dietary and occupational diversity in an under-developed region.

Greece

Greek mythology, archaeological research, and historical records all show that the economic, political, social, cultural and religious life of Greece has been interlinked with wetlands for the last 3,000 years. Springs, for example, were protected through religious taboos, rivers were feared for their potential flood damage, and marshes were considered as places of ill health and of little practical use. Throughout history, efforts were made to alter wetlands in order to prevent flood damage, provide drinking and irrigation water, and to acquire more farmland. However, these efforts had little or no effect on the country's wetland geography until 1922. At that time, following World War I, 1.5 million refugees, mostly farmers, poured into Greece from Asia Minor. To accommodate this massive immigration, the Greek Government, with the help of foreign aid, invested in the rapid development of the available soil and water resources to provide more farmland and irrigation water, and to eradicate malaria. As a direct consequence, most of Greece's marshes, and several lakes and lagoons, have been completely drained within the last two generations. Rivers and streams have been dammed, diked and rerouted, and floodplain forests felled. In total, some two-thirds of the wetland area of the country has been lost, with subsequent detrimental consequences to the country's biological diversity.

Conservation support

In spite of this destruction, Greece still possesses some of the Mediterranean's most significant wetlands, especially in Macedonia and Thrace, and along the western coast. In total, 263 wetland sites have been identified, all of which merit conservation action in the face of continuing agricultural, industrial and tourist development.

In response to the growing concern for wetlands in Greece, WWF and the European Commission helped the Ministry of Environment and the Goulandris Museum to set up a Greek Wetlands Centre. The centre was established in 1991 and is located in Salonika, the capital of Macedonia; an apt location as Salonika is close to some of Greece's most important wetlands, such as Lakes Mikri and Megali Prespa, and Lake Kerkini. An important role of the centre is to monitor the status of the country's wetlands. With the information acquired the goal is to build greater awareness among the government and the public of the pressures facing wetlands, and to encourage positive action to redress the situation.

Middle East

...sopotamia, the land between the Tigris and Euphrates rivers, is the ... site of the Garden of Eden. Here, in the largest continuous wetland system in the Middle East, civilization was established as early as 4,000 years BC, and a sophisticated irrigation system was developed. Today, in what is now southern Iraq, these marshes still cover 15,000 square kilometres (6,000 square miles), their luxuriant growth contrasting with the arid landscape of most of the Middle East.

This complex of shallow freshwater lakes and marshes, river channels and seasonally inundated floodplains owes its existence, together with most other wetlands of the region, to the mountain ranges of eastern Turkey, northern Iran and Afghanistan. There is sufficient rain- and snowfall to feed several major river systems, which in turn sustain several impressive wetlands, including the lakes of the Rift Valley in the Levant and Jordan, the vast marshes of the Tigris and Euphrates in Mesopotamia, the wetlands of the Orumiyeh Basin, South Caspian lowlands, Khuzestan and central Fars in Iran, and the wetlands of the Seistan Basin on the Iran/Afghanistan border.

Although the Mesopotamian marshes of southern Iraq are the most extensive wetland ecosystems in the Middle East, a similar but much smaller complex of wetlands lies in neighbouring southwestern Iran. Three large rivers, the Karun, Dez and Kharkeh, rise in the Zagros Mountains. These flow out onto the plains of Khuzestan, creating an area of seasonal floodplain wetlands, which extend south to the head of the Gulf. In northern Iran, the marshes of the South Caspian lowlands are equally impressive. Here there is an almost unbroken chain of freshwater lakes and marshes, brackish lagoons, irrigation ponds (locally known as "ab-bandans") and rice paddies stretching for over 700 kilometres (400 miles) along the Caspian shore. Fed by abundant rainfall on the humid north slope of the Alborz Mountains, most of these wetlands are permanent and support vegetation all year. The two largest wetlands in this region are the Anzali Mordab complex in the southwest Caspian, and the Gorgan Bay/Miankaleh Peninsula complex in the southeast.

Wetland diversity

Many of the Middle East's major rivers lie in internal drainage systems, ending up in land-locked, brackish to saline lakes which are often subject to wide fluctuations in water level and may dry out completely during periods of drought. Extensive fresh to brackish marshes occur where rivers and spring-fed streams enter several of these salt lakes. Sabkat al Jabboul in Syria, the Dead Sea in the Jordan Valley, Lake Orumiyeh, Gavkhouni Lake and the Neiriz Lakes in Iran, and the wetlands of the Seistan Basin are all examples. The latter are unusual in that although the three main lakes of the Seistan Basin, namely Hamoun-i Puzak, Hamoun-i Sabari and Hamoun-i Helmand, lie within an internal drainage basin, they are predominantly freshwater. The system is fed by the Helmand River, which rises in the Hindu Kush in northern Afghanistan. In most years, the Helmand supplies sufficient water to flood only one or two of the lakes, but about once every decade, the floodwaters of the Helmand sweep through all three lakes and overflow into a vast salt waste to the southeast, flushing the salts out of the system.

Further south in the Middle East, there are very few natural wetlands away from the coast. Azraq Oasis, in Jordan's Eastern Desert, is a notable exception, but this has now been reduced greatly by uncontrolled extraction of groundwater. In the extremely arid interior of the Arabian Peninsula, the only significant natural wetlands are the *playa* wetlands (or "qa") of northern Saudi Arabia – low-lying desert areas which are temporarily flooded during periods of heavy winter rainfall. In some parts of Yemen, sub-surface seepage feeds saline grazing marshes in valley bottoms, while in southwestern Oman, the slight influence of the southwest monsoon provides sufficient rainfall to maintain a chain of brackish creeks and lagoons along the narrow coastal plain.

KAZAKHSTAN

Volga Delta

*Aral
Sea*

RUSSIAN
FED.

*Caspian
Sea*

UZBEKISTAN

Ramsar Sites
Parks and Reserves
Water Bodies
Mangrove
Estuaries, Coastal Wetlands
Soda Lakes
Freshwater Marsh, Floodplains

GEORGIA
Tbilisi •

AZERBAIJAN
Baku •

ARMENIA
Yerevan •

Krasnovodsk and
North Cheleken Bays

TURKMENISTAN

Kirov Bays

*Orumiyeh
Basin*

Lake
Oroomiyeh

• Lake Gori

Anzali Mordab Complex
Bandar Kiashahr Lagoon
and mouth of Sefid Rud
Amirkelayeh Lake

Alagol, Ulmagol
and Ajigol Lakes

• Ashkhabad

• Lake Kobi

Shur Gol, Yadegarlu &
Dorgeh Sangi Lakes

Alborz Mountains

Miankaleh Peninsula, Gorgan Bay
and Lapoo-Zaghmarz Ab-bandans

• Tehran

Middle East Visitors

Tigris

Mesopotamia

Zagros

IRAN

AFGHANISTAN

*Seistan
Basin*

Many millions of
waterbirds, notably grebes,
cormorants, ducks, geese,
swans, cranes, coots,
shorebirds, gulls and terns,
which breed in Siberia,
spend the winter at
wetlands in the Middle
East or use these as
refuelling areas on their
way to and from wintering
areas in Africa or the
Indian Sub-continent. The
marshes of Mesopotamia
and the South Caspian
lowlands are particularly
important for ducks and
geese (over 20 species),
while the mud flats of
the Gulf and Arabian
Sea coasts are of critical
importance for shorebirds,
gulls and terns.

Dez
Karkheh

Karun

Gavkhouni Lake and marshes
of the lower Zaindeh Rud

Hamoun-e-Saberi

South end of
Hamoun-e-Puzak

Baghdad

KHUZESTAN

RAQ

Basra •

KUWAIT

Mountains

Neiriz Lakes and
Kamjan Marshes

Lake Parishan and Dasht-e-Arjan

Helmand

PAKISTAN

Shadegan Marshes and
tidal mud flats of Khor-al
Amaya and KhorMusa

FARS

*Tarut
Bay*

*Hara
Protected
Area*

Khuran Straits

Deltas of Rud-e-Shur,
Rud-e-Shirin
and Rud-e-Minab

Dammam Port •

• Manama

*The
Gulf*

Deltas of Rud-e-Gaz
and Rud-e-Hara

QATAR
• Doha

• Riyadh

Gulf of Oman

| 0 km | | 500 | | 1,000 |

| 0 mile | | 250 | | 500 |

SAUDI ARABIA

U.A.E

• Abu Dhabi

• Muscat

OMAN

*Barr-al-
Hekman*

By contrast, the coastline of southern Iran and the
Arabian Peninsula possesses extensive wetlands.
Intertidal mud flats run along the coasts of the Gulf
and southern Red Sea, and also in some areas of
the Arabian Sea coast. They are very extensive at
the head of the Gulf in Iran, Iraq and Kuwait,
along the southern shore of the Gulf in Saudi
Arabia and the Gulf States, at Barr-al-Hekman on
the Oman coast. Many of these mud flats support
simple mangrove communities consisting of only a
single species, *Avicennia marina*. Much the largest stands of mangroves grow in
the Khuran Straits in southern Iran, where there are some 1,000 square
kilometres (400 square miles) of low-lying islands with broad mangrove
fringes and extensive intertidal mud flats.

YEMEN

• Sanaa

• Ta'izz

Competition for Water

The wetlands of the Middle East have suffered from a range of human impacts, including drainage, pollution and in-fill for urban and industrial development. However, in a region in which water is a scarce resource almost everywhere, wetlands have come under particularly severe pressure. Throughout Iran, Iraq, Syria, Lebanon, Israel and Jordan, water supplies have been diverted for irrigation purposes, and domestic and industrial consumption, while the wetlands themselves have been drained for agriculture, industry and urban development. The increasing utilization of the waters of the Tigris and Euphrates rivers for irrigation in Turkey, Syria and upper Iraq has resulted in a significant loss of wetlands in Mesopotamia. Flood control projects and irrigation schemes on the Helmand River in Afghanistan have also had an adverse effect on the wetlands of the Seistan Basin in Iran, especially during years of below-average rainfall.

In Syria, Lebanon and Israel, almost all of the original freshwater wetlands were drained for agriculture at the beginning of the 1900s, while in the Jordan Valley, the level of the Dead Sea has fallen dramatically as a result of water being diverted from the Jordan River for irrigation schemes.

The Euphrates under threat

The largest scheme currently being developed in the region is an ambitious project to harness the Euphrates to produce power and irrigate over 185 square kilometres (70 square miles) of arid land in Turkey's southeast provinces. However, full development of the scheme could mean that Syria would lose 40 per cent of its Euphrates water and Iraq up to 90 per cent. Already, Turkey has had to close off the Euphrates for one month (in January 1990) to begin filling the reservoir behind the Ataturk Dam. Although this is the most important dam, it is only one of 21 that are being planned in an attempt to irrigate an area of southeast Turkey. The area under discussion covers some 108,000 square kilometres (42,000 square miles), a land area the size of Cuba.

The Marsh Arabs

Undoubtedly the best-known inhabitants of the wetlands of lower Iraq are the Ma'dan or marsh Arabs. Made famous by the tales of the English explorer Wilfred Thesiger, the Ma'dan have lived in these marshes for over 5,000 years, isolated from the rest of Iraq by the extensive wetlands. For the Ma'dan, whose isolation has resulted in a culture that has changed little over hundreds of years, the lakes and marshes that are prevalent in southern Iraq are essential to their livelihoods. The expansive reedbeds, for example, provide the Ma'dan with their staple building material, from which they construct their boats and artificial island houses, while the lakes on which the houses float have always been important fishing grounds.

Formerly spear-fishermen, catching species of barbel and carp for their own needs, the Ma'dan have recently taken to using nets to improve their catch for export to Basra and Baghdad. Mat-weaving has also become an important source of income, as demand for the versatile coverings, which are used in packaging and fencing, as well as a building material, has increased. However, it is buffalo that have remained the basis of family wealth. As well as providing the Ma'dan with meat and hides, which are used to make a variety of products, the herds of buffalo also generate milk, which is turned into butter and yogurt. In addition, today the Ma'dan also cultivate rice to supplement their diets.

A threatened existence

Following the Gulf War of 1990/91, the Ma'dan suffered from continuing violence in the region. Their villages were attacked and bombed by helicopter gunships. While the Ma'dan people themselves are threatened by war, their marsh homeland is also under threat from irrigation and oil exploration, as well as a major water diversion project, known as the Three River Project. If the project goes ahead, it could result in the draining of large areas of the marshes.

Top left **Ma'dan women gathering and stacking reeds near Al Kabayish at the southern edge of the Iraqi marshlands.** After drying, the reeds will be put to a variety of uses, including matting, fencing and housing. Traditionally, bundles of dried reeds were also used to build the rafts and boats which are the only viable form of transport in this wetland region. Today, the Ma'dan prefer to use wooden or metal boats because they do not require annual replacement.

Above left **Traditional Ma'dan reed house.** The sides and roof are made of woven reed matting, and are supported on a framework which uses tightly bundled reeds as pillars and beams. The design of these houses has not changed in more than 5,000 years.

Left **Lumps of crystallized salt litter the surface of the Dead Sea in Israel.** The extreme salinity of this inland sea places it among those very few wetlands which are virtually lifeless.

Man-made Wetlands

The rapid urban and agricultural development in the Arabian Peninsula in recent decades has resulted in the creation of a large number of artificial wetlands. The majority are either water storage reservoirs, areas of spillage from irrigation systems, sewage treatment ponds or artificial lagoons created by waste water from urban and industrial areas. Some of these wetlands can be surprisingly large, and many rank among the most important freshwater wetland habitats for wildlife in the peninsula.

A number of the larger artificial wetlands, such as the reservoirs in Wadi Hanifah and Wadi Jizan in Saudi Arabia, support several thousand waterfowl during the winter months. In addition, sewage lagoons and irrigation ponds provide excellent stop-over areas for a wide variety of migrant birds.

Azraq Oasis

The wetlands of Azraq Oasis, in Jordan's Eastern Desert, lie at the heart of a large internal drainage basin. The basin extends from the Druze Highlands in Syria to the borders of Saudi Arabia. The oasis wetlands themselves formerly comprised about 15 square kilometres (6 square miles) of permanent areas of marshes and pools, which lay alongside a seasonally flooded *playa* wetland known as Qa Azraq.

Qa Azraq originally covered an area of about 60 square kilometres (23 square miles). It was, and indeed still is, an outstanding example of a seasonal wetland in a predominantly arid region. As well as supporting a rich and varied flora and fauna, the wetland is a vitally important "refuelling" area for huge numbers of migratory birds. Qa Araq was designated as a Ramsar Site in 1977.

A stretched resource

Since 1982, however, the spring-fed marshes have suffered drastically as a result of wide-scale groundwater extraction. The amount of water pumped to Amman, for example, has increased by 60 per cent, while there has also been a rapid proliferation in the number of wells extracting water for irrigation. By 1991, there were some 450 irrigation wells, mostly illegal, pumping an estimated 22 million cubic metres (790 million cubic feet) of groundwater per year. As a result, the water table has been lowered by between 6 and 10 metres (20 and 33 feet).

Three of the four main springs which feed the wetland have dried up, and the remaining spring is discharging at a fraction of its original rate. Perhaps of greater concern are signs that the upper aquifer, which, for many years, fed the wetlands' springs, is now being contaminated with saline water from the lower aquifers. It is feared that by the mid-1990s, the groundwater will no longer be suitable for human consumption. Increasing soil salinity, as a result of using slightly saline water for irrigation, will also render much of the area unsuitable for agriculture.

The *playa* wetland has been unaffected by the pumping, and continues to flood during periods of heavy winter rains. However, in July 1991 the Jordanian Government announced plans to construct a dam on Wadi Rajil, 45 kilometres (27 miles) north of the oasis, to provide water for irrigation. As Wadi Rajil is the single most important wadi (temporary river course) feeding flood waters into Qa Azraq, the dam is likely to have a serious impact on the extent of winter flooding in the *playa* wetland.

Unless some immediate action is taken at Azraq, one of the world's most outstanding oasis wetlands, which was formerly capable of supporting more than 100,000 wintering waterfowl and perhaps as many as a million migrant birds, will have been destroyed. Recognizing this crisis, the Government of Jordan is currently discussing support for Azraq from the United Nation's Global Environment Facility.

Below The pools and marshes at Azraq Oasis in Jordan represent the water-wealth of a vast arid area. In recent years, however, extraction of groundwater has lowered the water table, and the wetlands are now vastly reduced.

Bottom The spring thaw in the Zagros Mountains of Iran releases meltwater that flows into the complex wetlands of several valleys. Among the habitats created by the thaw are freshwater lakes and marshes, saline marshes and salt pans.

Right The Dalmatian pelican, which is also found in certain parts of southern Europe, recently has joined the list of globally endangered species. Because the Dalmatian pelican is superbly adapted for catching fish is often perceived as a direct competitor by humans, who are after the same prey. As a result, pelicans are often shot on sight, and some of their nesting colonies in the Middle East have been destroyed by irate fisherman.

The Sarmatic Fauna

Ten of the waterbirds most closely associated with the Middle East are believed to have evolved around the shores of the ancient Sarmatic Sea. About three million years ago, this vast saline inland sea extended as a continuation of the present eastern Mediterranean, through the Black Sea and Caspian Sea to the Aral Sea and beyond. As this sea became fragmented into a chain of smaller inland seas and salt lakes, the populations of waterbirds associated with it similarly became fragmented, and most of these species are now extremely scarce.

While a few of the species, such as the red-crested pochard (*Netta rufina*), great black-headed gull (*Larus ichthyaetus*) and slender-billed gull (*L. genei*) remain at least locally common, four species, namely the Dalmatian pelican (*Pelicanus crispus*), pygmy cormorant (*Phalacrocorax pygmeus*), marbled teal (*Marmaronetta angustirostris*) and white-headed duck (*Oxyura leucocephala*) are now considered at risk.

Conservation in the Middle East

Despite the significance of wetlands in arid regions, of all the countries in the Middle East, only Iran has made any serious attempt to conserve its dwindling wetland resources.

The first protected area incorporating a major wetland (Lake Orumiyeh) was established in 1967 by the Iran Game and Fish Department, later to become the Department of the Environment. By 1977, this department had established no less than 68 protected areas and refuges, covering over 78,000 square kilometres (30,000 square miles). Within this area, examples of all the country's major natural ecosystems were represented. This excellent system of reserves survived the political upheavals of the late 1970s almost intact, and by 1991 had been expanded to 77 reserves covering over 800,000 square kilometres (300,000 square miles). Wetlands figure prominently in this network of reserves. One of the national parks, eight of the protected areas and seven of the wildlife refuges were established primarily to protect wetland ecosystems, while a further five protected areas and two wildlife refuges incorporate important wetland habitats.

East Africa and the Nile Basin

The landscape of East Africa is dominated by three physical features: the Great Rift Valley, which runs through the eastern side of the continent, Lake Victoria, and the valley of the River Nile as it flows north to the Mediterranean. These physical features, together with the coastal systems of the Indian Ocean, Red Sea and the Mediterranean, support a variety of wetlands ranging from the vast expanse of the Sudd floodplain in southern Sudan to the glacial lakes of Mount Kilimanjaro.

The high-rainfall areas of East Africa surround Lake Victoria and provide the Nile with more than half its water. The entire surface of Uganda drains its waters into the Nile Basin, as do parts of Kenya, Tanzania, Rwanda and Burundi. This wet region supports several large and many small wetlands associated with its rivers and lakes, the largest being the edges of Lake Victoria, which extend through three countries. The more moderate rainfall of Tanzania feeds several large floodplain wetlands, the most notable being the Malagarasi/Moyowosi system, which drains into Lake Tanganyika, the Kilombero floodplain, which drains into the Rufiji River and Indian Ocean, and the Wembere floodplain, which drains into the Rift Valley.

Most of neighbouring Kenya is much drier. Its large mountains, Mount Kenya and Mount Elgon, give rise to rivers that cross the desert and develop important floodplain wetlands en route – especially the Tana River floodplain and delta. Several seasonal floodplains and swamps are present in dry northern Kenya, the two most important being the Lorian in the northeast and the Lotagipi in the northwest, which is fed by watersheds in neighbouring Sudan. The highest mountain in the region is Mount Kilimanjaro in Tanzania, which despite being so close to the Equator, has glaciers and melting snow that feed springs further down its slopes, producing swamps and floodplains.

The Nile and Horn of Africa

As the Nile flows north into the Sudan, it spreads out to form the Sudd floodplain. Covering 16,500 square kilometres (6,400 square miles) of permanent swamp and up to 15,000 square kilometres (5,800 square miles) of seasonal floodplain, the Sudd is one of the three largest wetlands in Africa, and the stronghold of the Dinka people. As it leaves the Sudd, the White Nile is joined by the Blue Nile, which drains the Ethiopian Highlands. In this mountainous country there is a diversity of wetlands ranging from the extensive salt marshes and isolated mangroves of the Red Sea coast and the lakes of the Rift Valley to the montane lakes and seasonal floodplains of the rivers draining the mountains. In Somalia, wetlands are confined largely to a few lagoons, mangroves and other tidal wetlands on the coast, and swamps and floodplains along the Jubba and Shebelle, the two large perennial rivers which cross the country from Ethiopia.

In Egypt, the Nile today is tightly constrained, first by the Aswan Dam and by canals, dikes and irrigation structures along its course to the Mediterranean. These changes have reduced the delta and floodplains of the lower Nile to a vestige of their former selves. The vast papyrus swamps, for example, which were characteristic of the delta during the ancient Egyptian civilization, have almost totally disappeared from the region.

Coastal wetlands

The main coastal wetlands lie in deltas and estuaries where extensive stands of mangroves have developed, particularly in the Tana Delta and Vanga-Funzi in Kenya, and the Pangani and Rufiji deltas in Tanzania. The mangroves are currently threatened by overharvesting for construction poles, firewood and charcoal, and timber for furniture. Clear-cutting for urban development, creation of prawn farms, expansion of port facilities, diversion of freshwater, oyster harvesting, and siltation from the upper catchment areas are also responsible for their degradation. In response, a series of conservation efforts

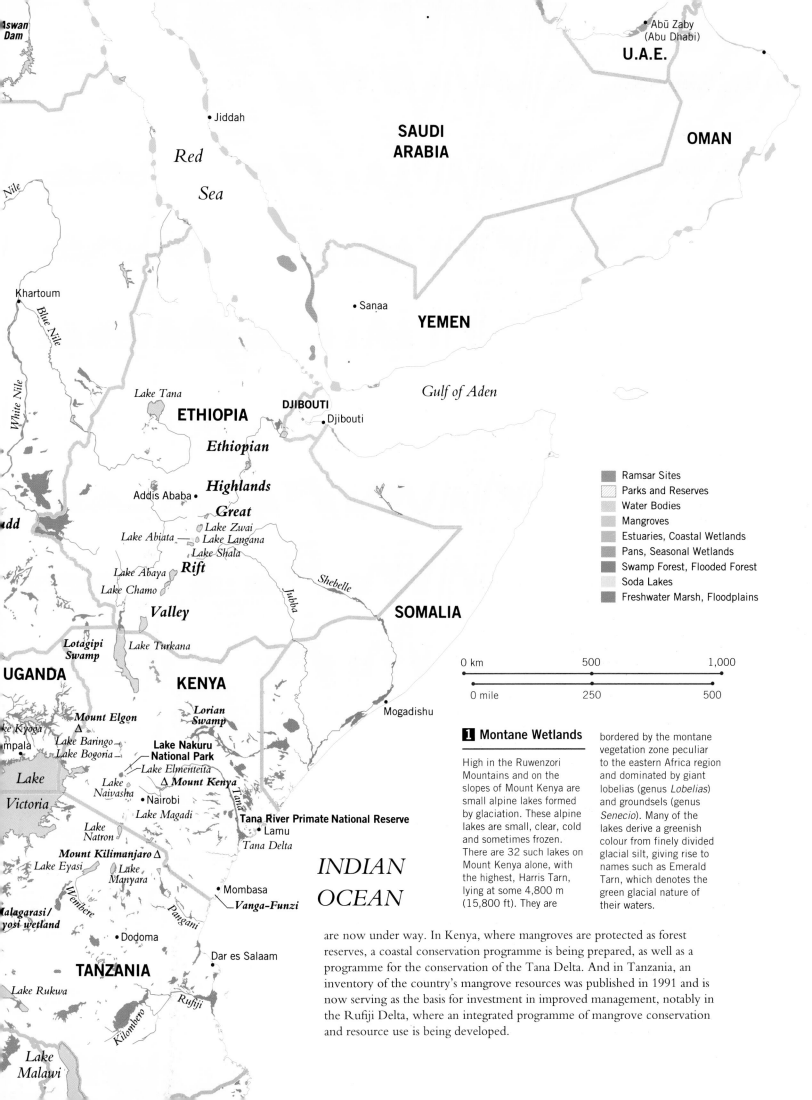

Aswan
Dam

Red
Sea

SAUDI
ARABIA

• Jiddah

• Abū Zaby
(Abu Dhabi)

U.A.E.

OMAN

Nile

Khartoum

Blue Nile

White Nile

dd

• Sanaa

YEMEN

Gulf of Aden

Lake Tana

ETHIOPIA

DJIBOUTI

• Djibouti

Ethiopian

Highlands

Addis Ababa •

Great

Lake Zwai
Lake Abiata — Lake Langana
Lake Shala

Rift

Lake Abaya
Lake Chamo

Valley

Shebelle

Jubba

SOMALIA

Lake Turkana

Lotagipi
Swamp

UGANDA

KENYA

Lorian
Swamp

Mogadishu

ke Kyoga
mpala

Mount Elgon
△

Lake Baringo
Lake Bogoria

Lake Nakuru
National Park

Lake Elmenteita

△ Mount Kenya

Lake
Naivasha

• Nairobi

Lake Magadi

Tana

Tana River Primate National Reserve

• Lamu

Tana Delta

Lake
Victoria

Lake
Natron

Lake Eyasi

Mount Kilimanjaro △

Lake
Manyara

Wembere

INDIAN
OCEAN

Malagarasi/
yosi wetland

• Mombasa

Vanga-Funzi

Pangani

• Dodoma

Dar es Salaam

TANZANIA

Lake Rukwa

Rufiji

Kilombero

Lake
Malawi

Ramsar Sites
Parks and Reserves
Water Bodies
Mangroves
Estuaries, Coastal Wetlands
Pans, Seasonal Wetlands
Swamp Forest, Flooded Forest
Soda Lakes
Freshwater Marsh, Floodplains

0 km   500   1,000

0 mile   250   500

1 Montane Wetlands

High in the Ruwenzori
Mountains and on the
slopes of Mount Kenya are
small alpine lakes formed
by glaciation. These alpine
lakes are small, clear, cold
and sometimes frozen.
There are 32 such lakes on
Mount Kenya alone, with
the highest, Harris Tarn,
lying at some 4,800 m
(15,800 ft). They are

bordered by the montane
vegetation zone peculiar
to the eastern Africa region
and dominated by giant
lobelias (genus *Lobelias*)
and groundsels (genus
Senecio). Many of the
lakes derive a greenish
colour from finely divided
glacial silt, giving rise to
names such as Emerald
Tarn, which denotes the
green glacial nature of
their waters.

are now under way. In Kenya, where mangroves are protected as forest
reserves, a coastal conservation programme is being prepared, as well as a
programme for the conservation of the Tana Delta. And in Tanzania, an
inventory of the country's mangrove resources was published in 1991 and is
now serving as the basis for investment in improved management, notably in
the Rufiji Delta, where an integrated programme of mangrove conservation
and resource use is being developed.

Lakes of the Rift Valley

Stretching from Lebanon to Mozambique, the Great Eastern Rift Valley is a 6,500-kilometre (4,000-mile) fissure in the Earth's crust. In East Africa, it is at its most dramatic and provides the foundation for a series of lakes which are among the most exotic features of the region's wetland diversity. There are three major types of lakes: freshwater, moderately saline and hypersaline.

In Ethiopia, there are six lakes strung along the Rift Valley – Zwai, Langana, Abiata, Shala, Abaya and Chamo. In Kenya, the line is picked up and continued southwards through a chain beginning with the biggest and moderately saline Lake Turkana, then continuing along the valley with lakes Baringo, Borgoria, Nakuru, Elmenteita, Naivasha and Magadi. Again, these lakes vary from freshwater to hypersaline. In Tanzania, the chain continues with lakes Natron, Manyara, Eyasi and Rukwa; while in Malawi, Lake Malawi also lies in the Rift Valley.

The western arm of the Rift Valley includes lakes Kivu, Edward, Albert and Tanganyika. Lake Tanganyika, with a maximum depth of 1,470 metres (4,850 feet), is the second deepest lake in the world after Lake Baikal in Siberia. The lake is home to over 500 animals that are unique to the area, most of which evolved within the lake basin itself.

The most serious immediate problems facing the Rift Valley lakes have come about from overpopulation within the lakes' basins. Excessive suspended sediment inputs into the lakes caused by deforestation and overgrazing,

pollution, eutrophication, introduction of exotic species and overharvesting of the fragile aquatic resources have resulted in serious reductions of species and localized extinctions.

Soda lakes

Several of the lakes of the Rift Valley have exceptionally high concentrations of salts, and are known as soda lakes. Their waters are usually thick, slimy and greenish-looking because of the dense population of phytoplankton. Fish, however, are absent from most of the soda lakes, although one small species, *Tilapia alcalica grahami,* is able to tolerate the highly alkaline water of Lake Magadi in Kenya. This particular species of fish was also introduced to Lake Nakuru (also in Kenya) with great success in 1961, and has encouraged increasing numbers and varieties of waterbird to the lake to feed.

Soda lakes are also famous for their large concentrations of birdlife, and some such as Lake Nakuru, which supports over a million lesser flamingos at any one time, have been declared national parks.

Top left **Polygonal plates of crusted soda (sodium carbonate) cover the surface of Lake Magadi in Kenya. The soda erupts from underwater geysers in the middle of the lake. Apart from a few alkali-tolerant fish, the only inhabitants of the lake are billions of iron oxide-fixing bacteria. At times, the iron oxide stains areas of the lake a deep blood-red.**

Above **Not all of the lakes in the Great Rift Valley are hypersaline and lifeless. Further north, Lake Turkana supports many local fishermen and their families. The fish are caught in bowl-shaped baskets, and are then dried to preserve them.**

Left **Beneath a storm-shadowed sky, a shaft of sunlight illuminates some of Lake Nakuru's million lesser flamingos. The beak of the flamingo has a highly evolved filtration system, which enables it to ingest blue-green algae without also taking in the strongly alkaline waters of the lake.**

Tana Delta Wetlands

The Tana River Delta covers an area of approximately 1,300 square kilometres (500 square miles) and is the largest delta ecosystem in Kenya. Its riverine forests support two critically endangered and endemic species of primates, the Tana River red colobus monkey (*Procolobus rufomitratus*) and the Tana River crested mangabey (*Cercocebus galeritus*), while the 670 square kilometres (260 square miles) of floodplain grasslands, and associated woodlands and bushlands, lakes, mangroves, sand dunes and coastal waters, form a complex system which is unique on the Indian Ocean coast of Africa. Traditional land use practices of small-scale agriculture, pastoralism and fishing have maintained the ecological balance of the delta for thousands of years. The delta provides dry-season grazing and during droughts supports about 300,000 cattle. It also maintains vast numbers of wild herbivores; nearly 10,000 topi (*Damaliscus lunatus*), thousands of waterbucks (*Kobus ellipsiprymnus*), hippos and about 100 elephants.

Despite the many benefits to the local population and wildlife, the delta is currently threatened by a huge rice irrigation project which is expected to convert an area of 160 square kilometres (60 square miles) into a large-scale, highly mechanized rice farm. In response, a National Delta Wetlands Reserve has been proposed for the delta, and it is hoped that this will be Kenya's second Ramsar Site.

The Nile Perch and Lake Victoria

In 1960, an attempt was made to increase the fisheries of Lake Victoria by introducing the Nile perch (*Lates niloticus*), a species common enough in the Nile but not previously known in Lake Victoria. Following its establishment in the lake, the population of this large predator rose rapidly during the late 1970s and early 1980s, leading, it is believed, to a 10,000-fold reduction in the numbers of native fish which form the main diet of the Nile perch. These dramatic population declines have resulted in the extinction of many of the lake's rich fauna of over 170 cichlid fish species, 98 per cent of which were unique to the lake.

The Sudd, Dinka and Jonglei Canal

To many people, southern Sudan is a land of drought and civil war. Every year, the world's media brings us images of refugees fleeing by the thousand and famine on a massive scale. Yet in the midst of this human disaster lies the largest wetland system in East Africa. The Sudd, the Arabic word for "obstacle", is a vast floodplain where the White Nile leaves Uganda and flows north. Beginning 160 kilometres (100 miles) from Juba, and extending for some 540 kilometres (340 miles) to Malakal, the Sudd is a maze of thick, emergent aquatic vegetation. This vast wetland system is home for the Dinka, Nuer and Shilluk tribes, who have adapted their lives to the seasonal cycle of flood and drought.

When floods reach a peak in the latter half of the wet season, people congregate on high ground, bringing their cattle to graze on the upland grasslands that flourish with the July to October rains. At the same time, fields are cultivated, weeded and protected from pests, and cattle are herded in ever-increasing proximity as the countryside is transformed from an arid plain into a shallow inland sea. When the floods recede, the cattle are herded onto the floodplain or "toic". Here, seasonal inundation produces a rich cover of aquatic grasses, which remain for the dry season. Wildlife also concentrates on the floodplain during this period and hunting provides an important source of food.

Beyond the floodplain lies the permanently flooded swamp characterized by papyrus, bulrushes and open water. Here, specialist fisher-people live in villages on the banks of the main river channel, or on mounds that have been built up from detritus over the years. These people subsist largely on fish and what other foodstuffs they can trade for fish.

The Jonglei Canal

Egypt, Sudan's northern neighbour, has long viewed the waters of the Nile as the guarantor of food. However, as demands have increased, so the need to draw from the Nile has risen. In the latter half of the 1900s, special attention has focused upon the proposed Jonglei Canal. If completed, this canal would stretch 360 kilometres (210 miles) to bypass the Sudd and direct downstream some of the water that is lost from the river each year through spill and evaporation in the swamp. However, it is these "losses" which help yield the pasture, crops and fishery resources that are the basis of the local economy. The proposed diversion of the river has given rise to concern that the canal will destroy the livelihood of the peoples of the Sudd.

In the middle of much international attention, work on the Jonglei Canal started in 1978. However, in May 1983 a mutiny of troops on the southern edge of the Sudd marked the start of civil war in the area and all canal-related development had to stop. Today, it is war in the region rather than the canal that disrupts the lives of the people of the Sudd. But peace will bring with it a need for careful reappraisal of the Jonglei project and its effects on the people of the region.

Uganda's Response

For centuries, Uganda's wetlands have been important resources for the country's inhabitants. Today, with rising population and increased pressure upon dryland resources, Uganda's wetlands are of even greater significance. At least 10 per cent of Uganda is made up of swamps and floodplains, and the edges of the main lakes – Victoria, Kyoga, Albert, Edward and George – are of prime importance. Wetlands are, therefore, a national feature; they provide food, building materials (in the form of reeds) and water.

Since the latter half of the 1900s, Uganda's wetlands have come under increasing pressure from drainage for extension of pasture and crops, for rice paddy development or for expansion of dry land for other development

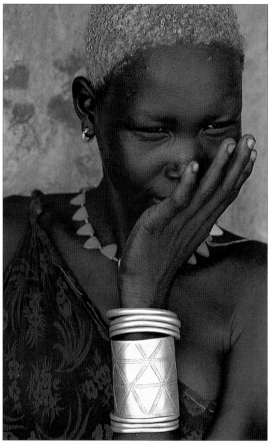

Above **Woman of the Nuer tribe pictured at home. When travelling, the semi-nomadic Nuer usually cover their bodies with ashes to protect themselves from insects.**

Top right **A Dinka man sings to his favourite bull to please it. The Dinka are completely dependent on their cattle. Cowhide is used for beds, bowls and drums; cooking utensils are carved from the bones, and the tails are made into talismans for young girls. The flesh is only consumed at rituals and celebrations. Drinking milk, or eating clotted blood, is the more usual way of obtaining protein from the cattle.**

Right **In Uganda's Queen Elizabeth National Park, the wildlife is protected. In other parts of the country, hippos are persecuted because they occasionally damage crops. The richness of the park is underlined by the large number of fish-eating cormorants and pelicans.**

projects. Agricultural and industrial pollution, watershed mismanagement, land use and land tenure disputes, invasion of aquatic weeds and over-use of resources are also causing wetland degradation. Swamp products such as papyrus are being overharvested, vegetation overgrazed, and the capacity of some wetlands to purify domestic and urban sewage, and agricultural run-off has become severely overstretched.

When President Yuweri Museveni came to power in Uganda in 1986, he brought with him, from time spent in the southwest of the country, personal experience of the importance of wetlands to rural people and of the problems they face. As a result, one of the first actions of the new Government was to create a Ministry of Environment Protection. The ministry set to work developing a national policy to protect, conserve and manage wetlands, thus giving wetlands the same attention as forests, agriculture and health, for which policies had already been prepared.

Showing the way

By assessing wetlands and their values, and consulting villagers, the Ministry of Environment Protection is steadily working towards the completion of its wetland policy. Already, the Government's success has made Uganda the leader in the field of wetland conservation in East Africa. Among the most effective ways of achieving this success has been the dialogue with local people through district councils. The talks ensured that the policy reflected the villagers' needs, as well as providing the ministry with important advice from the villagers on how Uganda should view and manage these valuable resources. Through this process of dialogue, the policy will represent as far as possible what Ugandans feel about their wetlands and how they should be managed, who owns them and who decides on what developments can or cannot occur in these areas.

West and Central Africa

West and Central Africa is a region of contrasts. It stretches from the peninsula of Cap Vert in Senegal to eastern Chad, from the northern border of Mali to the Nigerian coast, and from the mountains of northern Chad to southern Zaire. Travellers in the 1700s found that the region contained some of the most spectacular and varied landscapes in Africa. Within the area, for example, lie the desert of Mauritania, Mali and Niger, the Sahelian "desert edge" and the savanna and extensive tropical forest of the south.

Today, although population increase and other pressures of the last 80 years or so have caused substantial loss and degradation of natural ecosystems, many remain intact and are of central importance in the lives of the people of the region. Of the many various ecosystems, wetlands are among the most important, and remain central to the regional economy.

Floodplains and flooding

It is one of the ironies of the African landscape that three of the continent's most important wetlands, namely the floodplains of the Senegal, Niger and Chad river basins, are found in the semi-arid region of the Sahel. The Senegal and Niger rivers rise in the Fouta Djallon massif of Guinea, where the wet season can last for seven to nine months and annual rainfall can reach 3,000–5,000 millimetres (120–200 inches). The rivers then continue north through the wooded savanna and into the Sahel where, with a wet season of only three to four months, lakes and temporary marshes are the only wetlands of any significance beyond the floodplains. The area of each floodplain varies from month to month as the rivers flood and recede, and from year to year in accordance with rainfall variation in the catchment area.

Flooding is the driving force behind the productivity of the floodplains and of central importance to the role which wetlands play in the region. From as early as the 900s, these floodplains have been centres of population in an otherwise arid and thinly populated region. Today, the pastoral and agricultural economy of much of the Sahelian region is dependent upon the presence of these floodplains to provide dry-season grazing, a regular supply of freshwater for agriculture, and the principal source of animal protein, in the form of fish, for much of the year.

Coastal wetlands

Along the coast, rivers have led to the development of several extensive wetland systems over hundreds of thousands of years. Mauritania's coast is dominated by the Banc d'Arguin, a former estuary which today forms Africa's largest coastal wetland. To the south, the deltas of the Senegal, Saloum, Gambia and Casamance rivers add to the extent and diversity of these wetlands. The Bijagos Archipelago, which millions of years ago was once the delta of the Rio Geba and the Rio Grande de Buba in Guinea-Bissau, complements the Banc d'Arguin with its mosaic of palm-covered islands and extensive areas of mangrove.

The mangroves form part of a band of 8,500 square kilometres (3,300 square miles) from Senegal to Sierra Leone, while the Niger Delta in Nigeria supports an additional 10,000 square kilometres (4,000 square miles) of mangroves. Together with the coastal lagoons of Ghana, the Ivory Coast, Benin and Togo, the Niger Delta dominates the Gulf of Guinea and plays a critical role in supporting the region's rich wildlife, which includes manatees, pygmy hippos (*Choeropsis liberiensis*), three species of crocodile, turtle, forest elephant and chimp.

The Zaire River system

The most extensive wetlands of the region extend across Zaire and Congo. Known as the Zaire Swamps, they cover over 200,000 square kilometres (80,000 square miles) of the Zaire Basin, and are composed of swamp forest

WESTERN SAHARA

Cap Blanc
Baie d'Arguin
Banc d'Arguin National Park
Tidra Island

MAURITANIA

Cap Timiris

Nouakachott

Lake d'Aleg

Djoudj National Park
Rosso
Lake Rhiz
Richard Toll
St Louis
Lac de Guiers
Bassin du Ndiael

Gueumbeul
Bakel

Cap Vert
Dakar
Saloum
SENEGAL
Kayes

MALI

Delta du Saloum

Banjul
GAMBIA
Casamance
GUINEA-BISSAU
Bissau
Geba
Grande de Buba
Lagoa de Cufada
Bamako
Fouta Djallon

Bijagos Archipelago
GUINEA

Conakry

SIERRA LEONE
Freetown
Mount Nimba Δ
Sassandra

LIBERIA
Monrovia

Ramsar Sites
Parks and Reserves
Water Bodies
Mangroves
Estuaries, Coastal Wetlands
Pans, Seasonal Wetlands
Swamp Forest, Flooded Forest
Freshwater Marsh, Floodplains

1 Lake Chad

Spreading across the borders of Niger, Nigeria, Cameroon and Chad, Lake Chad, rarely more than 7 m (23 ft) deep, can cover 25,000 sq km (10,000 sq miles) when river flow is high, but recede to less than 10,000 sq km (4,000 sq miles) in times of drought. Ninety-five per cent of the lake's water comes from the Chari and Logone rivers, which rise in the mountains of southern Chad, Cameroon and Central Africa. The remaining 5 per cent come from diverse, intermittent rivers, in particular the El Beid, the Komadougou-Yobe, and the Yetseram, which rise in Nigeria.

2 Lake Volta

Lake Volta is the largest artificial lake in Africa. Created in 1964 by the closure of the Akosombo Dam 100 km (60 miles) upstream from the mouth of the Volta, the reservoir covers 850,000 sq km (330,000 sq miles) and stretches 320 km (190 miles) from north to south and 400 km (250 miles) from east to west. Built to provide hydroelectricity, the dam has rarely achieved full capacity as it takes 3 or 4 years of consecutive flood to fill the lake to its full capacity. The construction of the dam led to a substantial reduction in the fisheries of the Volta Delta.

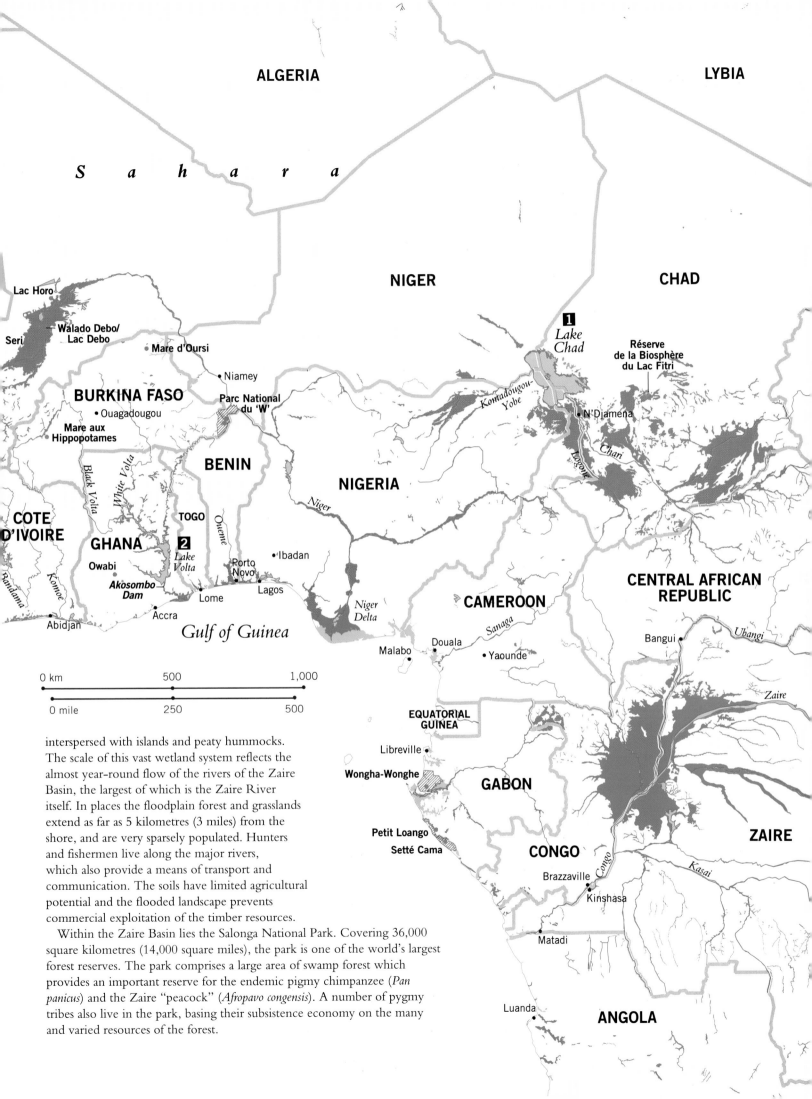

ALGERIA

LYBIA

S a h a r a

NIGER

CHAD

Lac Horo

Walado Debo/
Lac Debo

Seri

Mare d'Oursi

1 *Lake Chad*

Réserve
de la Biosphère
du Lac Fitri

Niamey

Komadougou-Yobe

BURKINA FASO

N'Djamena

Parc National
du 'W'

Ouagadougou

Logone

Chari

Mare aux
Hippopotames

BENIN

NIGERIA

Black Volta

White Volta

Niger

**COTE
D'IVOIRE**

TOGO

2
Lake Volta

GHANA

Owabi

Oueme

Porto
Novo

Ibadan

**CENTRAL AFRICAN
REPUBLIC**

Randama

Akosombo Dam

Lome

Lagos

Komoe

Accra

Niger Delta

Bangui

Ubangi

Abidjan

Gulf of Guinea

CAMEROON

Sanaga

Malabo

Douala

Zaire

0 km ———— 500 ———— 1,000

Yaounde

0 mile ———— 250 ———— 500

**EQUATORIAL
GUINEA**

Libreville

interspersed with islands and peaty hummocks.
The scale of this vast wetland system reflects the
almost year-round flow of the rivers of the Zaire
Basin, the largest of which is the Zaire River
itself. In places the floodplain forest and grasslands
extend as far as 5 kilometres (3 miles) from the
shore, and are very sparsely populated. Hunters
and fishermen live along the major rivers,
which also provide a means of transport and
communication. The soils have limited agricultural
potential and the flooded landscape prevents
commercial exploitation of the timber resources.

Wongha-Wonghe

GABON

ZAIRE

Petit Loango

Setté Cama

CONGO

Kasai

Brazzaville

Congo

Kinshasa

Matadi

Within the Zaire Basin lies the Salonga National Park. Covering 36,000
square kilometres (14,000 square miles), the park is one of the world's largest
forest reserves. The park comprises a large area of swamp forest which
provides an important reserve for the endemic pigmy chimpanzee (*Pan
panicus*) and the Zaire "peacock" (*Afropavo congensis*). A number of pygmy
tribes also live in the park, basing their subsistence economy on the many
and varied resources of the forest.

Luanda

ANGOLA

The Banc d'Arguin

The Banc d'Arguin is the largest intertidal wetland system in Africa. Lying between the two headlands of Cap Blanc to the north and Cap Timiris to the south, the Banc d'Arguin is a vast expanse of intertidal flats and creeks, where beds of eel grass (*Zostera* sp.) and other habitats provide important breeding and nursery areas for fish and crustaceans.

In the Baie d'Arguin, and surrounding the island of Tidra, two archipelagos of sand and rock islands support large breeding colonies of 15 species of waterbird. Found among the 25,000–40,000 pairs, there are great white pelicans (*Pelicanus onocrotalus*), greater flamingos, European spoonbills (*Platalea leucorodia*), and several species of heron, egret and tern, including two endemic sub-species. During the European winter, the tidal flats surrounding these islands provide a rich feeding ground for more than 2.3 million shorebirds, including bar-tailed godwit (*Limosa lapponica*), dunlin (*Calidris alpina*) and both ringed and grey plovers (*Charadrius hiaticula* and *Pluvialis squatarola*).

In addition to this avian richness, the combined influence of the cold Canaries current from the north and the warm Guinean current from the south make the Banc d'Arguin a frontier zone. Here, many plant and animal species from northern Europe and Asia at the southern limit of their range mingle with Afrotropical species at their northern limit. For example, the mangrove *Avicennia africana* is the most northerly in Africa, while the grass *Spartina maritima* is at its most southerly limit. Similarly, while the monk seal (*Monachus monachus*) and the common porpoise (*Phocoena phocoena*) are at the southerly limit of their range, the Atlantic humpback dolphin (*Souza teuzsii*) is not found north of the Banc d'Arguin.

The rich fauna and flora diversity of the region prompted the establishment of the Banc d'Arguin National Park in 1976, and its subsequent inclusion in the Ramsar List of Wetlands of International Importance and listing as a World Heritage Site.

The Imraguen fishermen

Every November to January, the Imraguen people fish for migrating mullet in the shallow waters of the Banc d'Arguin. Using long poles to beat on the water, they attract dolphin, which drive the mullet inshore and into their nets. The Imraguen's close dependence upon the tidal system, and their carefully adjusted fishing techniques, have made them the principal guardians of the Banc d'Arguin in the face of increasing fishing pressure from foreign fishermen. At a time when revenue from offshore fishing is Mauritania's principal source of export earnings, the Imraguen's careful use of resources inshore protects the most important fish nursery in the country.

In 1986, the International Banc d'Arguin Foundation was created, the purpose of which was to assist in providing international support to the Banc d'Arguin National Park. Today, the foundation is working closely with the Imraguen fishermen in order to maintain their fishing boats and market their products. In this way the foundation is laying down the basis for long-term sustainable use of the region's natural resources.

Bijagos Archipelago

Formed by the prehistoric delta of the Rio Grande de Buba and the Rio Geba, the Bijagos Archipelago consists of 88 islands and islets distributed over 10,000 square kilometres (4,000 square miles). Here, the rainy season brings fresh water into the coastal zone, while coastal currents from north and south meet, making the delta region vulnerable yet at the same time biologically rich. Between the islands, extensive mud flats are drained by a network of canals and creeks as the tide recedes.

Today, the characteristic vegetation of the islands are the palm groves which have replaced the once extensive forest. However, the tidal areas remain relatively untouched, forming a unique mosaic of mangroves and tidal flats. Here, hippos have adapted to live in sea water and can be seen plodding along the beaches, while otters hunt for shellfish or wallow in the creeks together with manatees for which the archipelago forms one of the most important strongholds in the region. Two species of dolphin also live here, including the rare Guinean dolphin (*Souza teuzsii*). Reptiles include two species of crocodile (*Crocodylus cataphractus* and *C. niloticus*) and four species of marine turtle, including the green turtle for which the Bijagos Archipelago is the most important breeding site in West Africa.

The people of the Bijagos

The archipelago is inhabited almost exclusively by the Bijagos ethnic group, 25,000 of whom live year round on 20 islands, using another 20 at particular times of the year. The Bijagos are careful to maintain the palms from which they extract oil and palm wine; they also harvest shellfish from the mud flats and catch fish for daily consumption.

In pursuit of enhanced economic development, both government and private and foreign entrepreneurs have in recent years turned their attention to the archipelago. Increasingly, the region is being seen as a focal point for development, with particular potential in tourism and sport fishing. However, the fragility of the natural environment and the specific cultural heritage of the people of the Bijagos has given rise to concern that, unless it is carefully planned, development investment will have serious social and environmental consequences. To address this concern an integrated development plan is being prepared. Among the measures included in the plan is the establishment of a biosphere reserve which would give special protection to critical ecosystems while encouraging sustainable use of natural resources in other parts of the archipelago.

Far left **Imraguen fishermen unload a catch of migrating mullet from their nets, while their aquatic co-worker waits offshore. The working partnership between the Imraguen and dolphins has existed for centuries. Today, however, this fragile alliance is threatened by the incursions of commercial fishermen.**

Above left **Although mangroves and tidal flats fringe much of the Bijagos Archipelago in West Africa, in other places tropical forest and palm groves extend all the way to the top of the beach. The Bijagos Archipelago is an important wintering ground for migratory birds. About 1.5 million shorebirds visit the area annually.**

Left **Greater flamingos filter food from the mud in the shallow waters around the Banc d'Arguin in Mauritania, West Africa. Meeting a variety of requirements – breeding, wintering, or migratory stop-over – the extensive mud flats of the Banc d'Arguin provide food and shelter for about 4 million birds every year.**

The Inner Niger Delta

During the 1400s, the Empire of Mali stretched from the coast of modern-day Senegal across southern Mauritania and Mali to the border of what is today Niger. To the south it encompassed the catchments of the Senegal and Niger rivers in Guinea, from where gold was traded across the Sahara. At the centre of this empire lay the Inner Niger Delta, the largest floodplain in West Africa, and a stable source of food for the population. Today, the Inner Niger Delta continues to play a central role in the lives of the region's people.

When the river is in full flood, the delta and its associated wetlands extend over some 320,000 square kilometres (120,000 square miles). It forms an inland sea in an arid region, the rainfall of which in some parts can be as low as 200 millimetres (8 inches). Over half a million people depend upon the delta, exploiting its rich soils, pasture and fisheries. As the annual rains begin to fall at the end of the dry season in June and July, local herders move their cattle from the dry bed of the delta to the higher ground on its edges where the rain stimulates fresh pasture. Over the next few months the herders move between pastures, exchanging milk products for millet and other dryland cereals grown by local farmers.

By September, towards the end of the wet season, floodwaters from the highlands of neighbouring Guinea transform the delta into a green oasis, enabling farmers to cultivate floating rice and other crops along the mosaic of channels and lakes that form the delta. Shortly afterwards, the cattle are brought back into the delta, where they feed on the luxuriant growth of floating grass (*Echinocloa stagnina*) known locally as "bourgou". At this time, the delta can support over a million cattle and a million sheep and goats. In 1985, 18,000 cattle and 257,000 sheep and goats were exported from the delta for a total value of US$8 million. The dry season is also the peak fishing season. About 80,000 people depend upon this resource; in 1986 an estimated 61,000 tonnes of fish were caught.

A threatened delta

The productivity of the delta has been a cornerstone of modern Mali since independence in 1960. However, the droughts of the 1970s and 1980s, together with rising population and changing administrative practices, have combined to increase pressure upon the delta's resources. In response, a major investment has been made in the delta. However, many of the schemes, designed to increase cereal and cattle production, have done little to meet the needs of the rural poor.

In search of solutions, studies of the way in which people use the resources have helped identify alternative approaches. The studies revealed that extreme poverty, exacerbated by years of drought, has forced families to use resources to meet short-term needs in a manner that is not sustainable. In response, current conservation efforts are supporting household economies through grain banks and revolving funds in order to provide the necessary economic stability upon which individual households can plan long-term strategies. At the same time, technical assistance is guiding families towards sustainable agriculture and discouraging practices that degrade the environment. Additionally, new legislation which allows greater popular participation in natural resource management encourages villagers to use resources carefully by ensuring that it is they who reap the benefits.

Village management

In the delta, the flooded woodlands of *Acacia kirkii* are the breeding grounds of waterbirds such as herons and cormorants. Faeces and regurgitated food from the breeding colonies fertilize the waters and support an economically important fishery. However, this mutually beneficial relationship is threatened.

The traditional land-use system within the delta divided the area into discrete fishing grounds managed by villagers during the flood season. When the same areas dried out, the herding communities, represented by a village

Left **In parts of Nigeria the Niger meanders through lush tropical forest. The seasonal flooding has little impact – the riverside huts are permanent dwellings. Further upstream, in the Inner Niger Delta floodplain, the river flows through a much more arid landscape, where the floodplain supports 2 million cattle and goats.**

Below left **Wood is in short supply in the virtually treeless landscape of the Inner Delta region. It is too scarce to be used for housing when the river provides a ready source of reed matting.**

Below **Bustling river traffic near the town of Mopti in central Mali. In addition to the seasonal floods, upon which the economy of the region depends, the Niger also provides a permanent and freely accessible transport system.**

leader called the *dioro*, administered the pastures and controlled access to herders. With independence in 1960, the Government of Mali nationalized all land. The management of fishing, grazing and woodland was put in the hands of the Ministry of Natural Resources, thereby greatly weakening traditional systems of resource management which had functioned for hundreds of years.

Overgrazing

Today, a major problem concerns the goat herders who use the delta during the dry season. By buying a Cutting Permit at the Forestry Department, the goat herders gain permission to construct a thorn enclosure where the goats spend the night. Although fines are levied on those who cut live trees rich in foliage to feed the goats, the fines imposed are collective as the culprit is rarely caught red-handed. As all herders in the area contribute to the fine, the rational herder decides to cut trees since he will pay anyway. The result is predictable; treecutting to give goats access to foliage has reached levels likely to damage the future of the woodland both for goat grazing and for the waterbird colonies (and hence for part of the fishery). An IUCN project is seeking a solution by encouraging the Forestry Department's efforts to create "Forêts Villageoises", which are managed by local communities consisting of goat herders, fishermen and *dioros*.

An essential element has been the retention of the traditional control by *dioros* over grazing areas. Ownership of the forest is vested in the village, which has interests primarily in fishing and is willing to recognize the traditional authority of the Dioros. The number of goat herds is reduced in one case to 20, a move which is welcomed by the goat herders. By re-creating traditional control structures, there is now hope that woodlands in the Inner Niger Delta will be conserved for the benefit of fisheries and waterfowl.

Southern Africa

In a landscape of meadows and forest in northwest Zambia, the Central African Plateau gives rise to the Zambezi River. Flowing to the other side of the continental divide, the Luapula River drains to the Zaire. These two river systems, the Zambezi and Zaire, drive the freshwater wetlands over much of southern Africa – from the small freshwater marshes of their catchment areas, to the extensive swamps and floodplains of their valleys.

The Luapula drains the vast swamps and floodplains of the Bangweulu region and its associated lakes – Bangweulu, Mweru and Mweru-Wantipa in Zambia and Zaire. Another tributary of the Zaire, the Lualaba, on the other hand, forms a complex landscape of floodplains, swamps and lakes, including Lake Upemba, in southern Zaire. The Zambezi River and its tributaries receive most of their runoff from Zambia and eastern Angola. They drain the great floodplains of western Zambia and the Kafue Flats, as well as the smaller floodplains of the middle Zambezi, Cuanda, Chobe/Linyanti and Luangwa rivers. For centuries, these floodplains have been crucial to the rural economies of the region.

Nowhere is the importance of these wetlands more evident than along the Zambezi, where many communities still organize themselves around the seasonal river flood. As the Zambezi flows back into Zambia from Angola it spreads out along the floodplains of Barotseland in what is today Zambia's Western Province. Stretching over 250 kilometres (150 miles) in length and 80–100 kilometres (50–60 miles) wide, the floodplain is flooded each year as the Zambezi rises above its low banks and spreads out over the landscape. As the grasses grow to keep above the flood, the plain becomes an intricate mosaic of green and blue.

The peoples of the floodplain

The Lozi people, who it is believed settled on the floodplain during the 1650s, continue to adapt to the annual flood cycle. Each year, as the river floods, the head of the tribe – the *litunga* – leads his people from the floodplain island that is their dry-season home to the higher ground along the edge of the valley where they spend the flood season. Today, this migration continues as a rich and colourful ceremony known as the *kuomboka*, marking an important point in the year for the people of the floodplain. Following nights of preparation, the chief's launch, crewed by 120 rowers, and a smaller barge for his queen, leads a flotilla of small boats across the lush greenery of the floodplain to the higher ground along the side of the valley.

While not all wetlands match the grandeur of the Zambezi floodplain, this seasonal cycle is repeated across much of southern Africa, even in small freshwater marsh depressions, known as "dambos". Although tiny when compared to the extensive African flooplains, dambos, however, play a central role in agriculture and grazing in many countries. Much of the landscape of Malawi, Mozambique, Zambia, southern Angola, Zimbabwe, and parts of Tanzania is dotted by dambos. Dry dambos are characterized by rich aquatic grasses and are treeless, while wet dambos are occupied by evergreen forests. Dambos are especially important in times of drought and during the dry season, when they frequently provide the only available grazing land, and provide the best crop harvests.

Lakes of southern Africa

The largest lake in the region, Lake Tanganyika, has limited wetlands because of its particularly steep basin and shores: it is an inland drainage system with a small outlet to the Zaire River. Lake Malawi (Nyassa) is the other large natural lake with extensive wetlands on its edges and along the Shire River, which joins it to the Zambezi. Two large man-made lakes, Kariba and Cabora Bassa, impound the Zambezi and produce viable wetlands when their management allows.

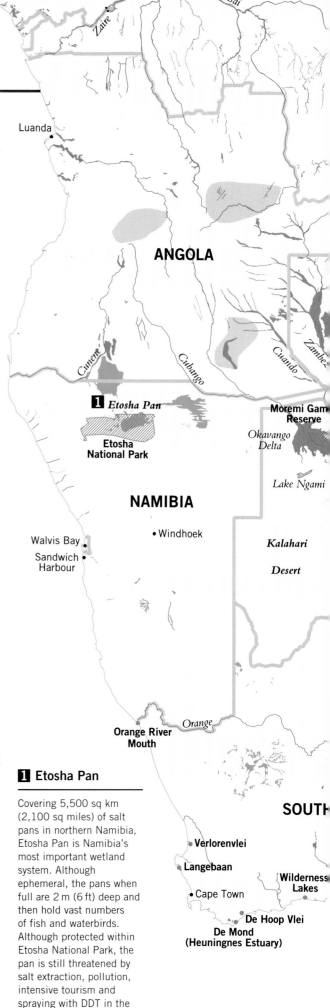

1 Etosha Pan

Covering 5,500 sq km (2,100 sq miles) of salt pans in northern Namibia, Etosha Pan is Namibia's most important wetland system. Although ephemeral, the pans when full are 2 m (6 ft) deep and then hold vast numbers of fish and waterbirds. Although protected within Etosha National Park, the pan is still threatened by salt extraction, pollution, intensive tourism and spraying with DDT in the catchment area.

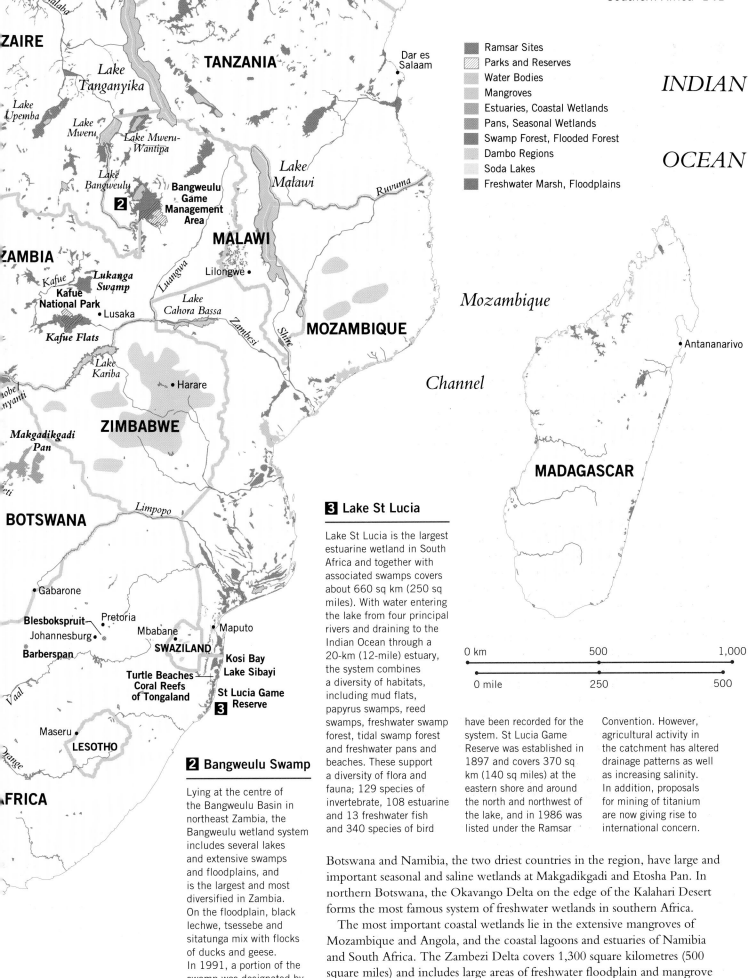

ZAIRE

TANZANIA

Lualaba

Lake Tanganyika

Lake Upemba

Lake Mweru

Lake Mweru-Wantipa

Lake Bangweulu

2

Bangweulu Game Management Area

MALAWI

Lake Malawi

Ruvuma

Dar es Salaam

ZAMBIA

Kafue

Lukanga Swamp

Luangwa

Lilongwe •

Kafue National Park

• Lusaka

Lake Cahora Bassa

Zambesi

Shire

MOZAMBIQUE

Kafue Flats

Lake Kariba

• Harare

Lake

ZIMBABWE

Makgadikgadi Pan

nyanti

obe

eti

BOTSWANA

Limpopo

• Gabarone

Blesbokspruit

Pretoria

• Johannesburg •

Mbabane •

• Maputo

Barberspan

SWAZILAND

Vaal

Turtle Beaches Coral Reefs of Tongaland

Kosi Bay Lake Sibayi

St Lucia Game Reserve

3

Maseru •

LESOTHO

Orange

FRICA

Mozambique

Channel

MADAGASCAR

• Antananarivo

INDIAN

OCEAN

	Ramsar Sites
	Parks and Reserves
	Water Bodies
	Mangroves
	Estuaries, Coastal Wetlands
	Pans, Seasonal Wetlands
	Swamp Forest, Flooded Forest
	Dambo Regions
	Soda Lakes
	Freshwater Marsh, Floodplains

0 km	500	1,000

0 mile	250	500

3 Lake St Lucia

Lake St Lucia is the largest estuarine wetland in South Africa and together with associated swamps covers about 660 sq km (250 sq miles). With water entering the lake from four principal rivers and draining to the Indian Ocean through a 20-km (12-mile) estuary, the system combines a diversity of habitats, including mud flats, papyrus swamps, reed swamps, freshwater swamp forest, tidal swamp forest and freshwater pans and beaches. These support a diversity of flora and fauna; 129 species of invertebrate, 108 estuarine and 13 freshwater fish and 340 species of bird have been recorded for the system. St Lucia Game Reserve was established in 1897 and covers 370 sq km (140 sq miles) at the eastern shore and around the north and northwest of the lake, and in 1986 was listed under the Ramsar Convention. However, agricultural activity in the catchment has altered drainage patterns as well as increasing salinity. In addition, proposals for mining of titanium are now giving rise to international concern.

2 Bangweulu Swamp

Lying at the centre of the Bangweulu Basin in northeast Zambia, the Bangweulu wetland system includes several lakes and extensive swamps and floodplains, and is the largest and most diversified in Zambia. On the floodplain, black lechwe, tsessebe and sitatunga mix with flocks of ducks and geese. In 1991, a portion of the swamp was designated by Zambia under the Ramsar Convention.

Botswana and Namibia, the two driest countries in the region, have large and important seasonal and saline wetlands at Makgadikgadi and Etosha Pan. In northern Botswana, the Okavango Delta on the edge of the Kalahari Desert forms the most famous system of freshwater wetlands in southern Africa.

The most important coastal wetlands lie in the extensive mangroves of Mozambique and Angola, and the coastal lagoons and estuaries of Namibia and South Africa. The Zambezi Delta covers 1,300 square kilometres (500 square miles) and includes large areas of freshwater floodplain and mangrove forest. In Namibia, the sheltered bays of the Namib coast contain important coastal wetlands such as those at Sandwich Harbour and Walvis Bay.

Wetland Antelopes

Many of Africa's wetlands are home to several of the continent's antelope species. The sitatunga (*Tragelaphus spekei*) is the most closely adapted to this aquatic life and although largely confined to the deepest, "impenetrable" swamps, is widespread across the continent. More specific to the wetlands of southern Africa are the antelopes of the genus Kobus. Waterbuck (*Kobus ellipsiprymnus*) live in the riverine wetlands and floodplains of most areas. A special floodplain and dambo species in southern and eastern Africa is the puku (*K. vardoni*) from Malawi, Zambia, Namibia and Botswana. However, the best-adapted wetland kob of the great floodplains and swamps of southern Africa is the lechwe (*K. leche*).

Lechwe are medium-sized antelopes with long, lyre-shaped horns and elongated toes that help them to walk in their muddy environment. They feed on floodplain grasses both on the edges of floods and in the water, often wandering up to a depth of a metre (3 feet). When disturbed, lechwe will usually flee deeper into the floodplains, using the water as a protection for their young, which are usually born on islands within the flooded grasslands or swamps. In the past, most populations of lechwe were prey to lions, which had learned to hunt them in the water. But in most places where lechwe live today, their main predator is man.

Lechwe distribution

Three subspecies of lechwe inhabit the floodplains and seasonal wetlands of Zaire, Zambia, Angola, Namibia and Botswana. In the past, the red lechwe (*Kobus leche leche*) was the most widespread. It ranged from southern Zaire and northern Zambia, to Angola and Botswana. Today, however, the red lechwe's most easterly home is the edge of Lukanga Swamp on the Kafue River in Zambia, where a small group shares its habitat with cattle. A protected population of red lechwe on Busanga Plain (in Zambia's Kafue National Park) presently numbers about 4,000 individuals, while scattered groups survive in the Barotse floodplains, the Cuando swamps and the Chobe/Linyanti floodplains close to the Zambezi. The largest group of red lechwe is found in the Okavango Delta of Botswana, where at least 20,000 animals inhabit the seasonally flooded grasslands.

The largest single population of kob is the Kafue lechwe (*K.l. kafuensis*), of which there are about 50,000 on the Kafue Flats in Zambia. Further north, the black lechwe (*K.l. smithemani*) has a smaller, widespread population in the Bangweulu wetlands and, possibly, is still present in neighbouring Zaire.

The Zambezi Basin

Far left **A confrontation between two male black lechwe raises splashes of water from the flooded surface of the Bangweulu wetlands, Zambia.**

Left **A pair of sitatunga browsing through aquatic vegetation. Sitatunga are found throughout sub-Saharan Africa, but are normally confined to the most isolated swamps. Like the lechwe, the sitatunga is well adapted to its wetland habitat. When frightened, sitatunga retreat into deeper water, submerging with only their noses above water.**

Below **The Zambezi River, upstream from the Victoria Falls. Once considered inexhaustible, the waters of the Zambezi are now the focus of regional concern. Downstream from the falls, the river feeds many lakes, and is heavily exploited for hydroelectricity. The possible diversion of water into the arid regions of northern Botswana, with consequent downstream water-loss, raises issues of international responsibility and cooperation. The obvious solution is "fair shares for all", but how are the size of the "shares" to be measured?**

The Zambezi River and its major tributaries flow through six countries in central and southern Africa – Angola, Botswana, Mozambique, Namibia, Zambia and Zimbabwe – while both Malawi and Tanzania also supply water to its basin. All these countries wish to use the Zambezi's water, particularly for its agricultural and hydroelectric potential.

Yet rainfall, water and wetlands, as well as the human populations that seek to exploit these and other resources, are unevenly distributed throughout the basin. As a result, the dependence of people on the Zambezi varies. For example, although Zambia contributes at least half of the water flowing to the Zambezi, the country is relatively wet and rarely critically dependent upon the river's waters for its agricultural and domestic supply. In contrast, Botswana contributes little to the river's flow, but views it as a potential solution to the water shortage in the north of the country. This combination of distribution and demand is a source of potential conflict.

In the mid-1980s, the United Nations Environment Programme (UNEP), together with the countries of the basin, embarked on an ambitious programme. The aim was to develop an environmentally conscious management system for the Zambezi River itself, as well as its watershed and its uses. This led to the production of the Zambezi Action Plan (ZACPLAN) in 1986 "for the environmentally sound management of the common Zambezi River system". ZACPLAN seeks to develop a network for monitoring water and water-development activities within the Zambezi Basin, and to work towards the cooperation of all states in water sharing and watershed management. Wetlands are included in the plan – not only as part of the watershed and its management, and as conservation areas, but also as sources of water and other resources for the region.

The Zambezi's essential wetlands

The issue of wetlands as sources of water is particularly important in southern and central African regions. Many countries are currently facing water shortages and are looking to new sources for their supply. Thus the wetland systems of the Okavango Delta, Chobe/Linyanti, Lake Kariba and Lake Cabora Bassa have all been considered as sources for "out-of-basin transfers" to alleviate water shortages – as has the flow of the Zambezi River itself. It is the intention of ZACPLAN to set in place a management and arbitration system that will ensure the sustainability of resource use in the Zambezi Basin; especially the use of vital water.

Ten years ago it was thought that the Zambezi's waters were infinite, a source for any who wanted them. Now it is clear that this is far from true. Watershed mismanagement has caused changes to the water quality and availability – both on the watershed and in the river itself – and extractions increase each year. ZACPLAN is therefore a timely programme for cooperation and monitoring. The hope is that, with extreme care, the Zambezi's resources can be shared in a way that ensures they will last.

The Okavango Delta

Lying on the northern edge of the Kalahari Desert, the Okavango Delta has for centuries provided an oasis for the tribes of northern Botswana, and a refuge for some of Africa's rarest birds and other wildlife. Today, with increasing pressure upon the country's water resources, the future of the delta is a major national and international concern.

Rising in the mountains of Angola, the Okavango River crosses into northern Botswana, where it flows for 100 kilometres (60 miles) along the "panhandle", a narrow valley between two geological fault lines, before spreading out to form the world's largest inland delta. From here to the downstream end of the delta, 97 per cent of the floodwater is lost through evaporation and transpiration. A small proportion spills over into the Boteti River and flows east to the Makgadikgadi pans (the largest salt pans in the world) or southwest to Lake Ngami.

The vegetation of the delta is rich and varied, ranging from water lilies and papyrus to floodplain forests. These habitats are home to a wealth of bird life, including little bee-eaters (*Merops pusillus*), jacanas (*Actophilornis africana*), malachite kingfishers (*Alcedo cristata*), grey herons, egrets and the remarkable African fish eagle (*Haliaeetus vocifer*), whose piercing cry is one of the most beautiful sounds of Africa's wetlands. The birds share the delta with over 15 species of antelope, including the shy sitatunga and the large herds of lechwe which splash across the floodplain. Hippopotamus, zebra and buffalo are all numerous, while Botswana's elephant population numbers over 60,000, the largest in Africa.

Right **Skilfully guided from the rear, a dugout canoe glides past a reedbed in the Okavango Delta region of Botswana. Instead of its traditional cargo – fish, game, or local produce – this canoe carries equipment and supplies for a tourist "safari". The profusion of wildlife has made the Okavango Delta a very popular tourist destination. Although most of the "safaris" are limited to photographic trophies, the Botswana Government does permit a limited amount of hunting.**

Tourists and the Okavango

People are naturally an integral part of the Okavango, and have lived in and around the delta for thousands of years. As well as introducing cattle, goats, sheep, horses and donkeys to the area, the people of the Okavango have also established an efficient agricultural system, growing maize, millet, sorghum and vegetables.

Today, a large proportion of the wetland is set aside as a wildlife preserve, Moremi Game Reserve, while an even larger area keeps large wild animals in and cattle out with a buffalo fence. Thus much of the central and upstream end of the delta is occupied by wildlife with a few villages and traditional hunters. This is the area favoured by the burgeoning tourist industry which brings visitors from the rest of the world. Many tourist operators have permanent or temporary "camps" in the wild parts of the delta. Photographic, wilderness tourism is combined with licensed safari hunting to form the main industry inside the delta. These activities are planned to continue under strict control so ensuring that it remains a sustainable industry that does not destroy its resource base.

A subsistence activity that recently has become commercial is the gathering of wetland products for village use and for sale in Maun, the largest town at the southern end of the delta. Water lily tubers, bulrush roots and palm hearts are collected for food, and palm wine is made from the sap of *Hyphaene* palms. But the most developed industry supplies fence, wall and roofing materials from reeds, sedges and grasses. Long stems of *Phragmites*

reeds are gathered in the delta and along the Boteti River, and then packed into bundles before being brought to town for sale. As long as the harvesting remains sustainable, the local population will be able to make a reasonable income indefinitely.

Urban and industrial growth

With a population of only around 1.3 million people, Botswana is one of the least densely populated countries in Africa. However, the population is growing at the rate of 3.5 per cent per year, and the demand for the Okavango's water, for use in agriculture, and for urban and industrial consumption, is high. Maun is already the largest town in the north of the country, and its predicted population increase will force water requirements to grow threefold by the year 2010.

However, this would still be small in comparison to the amount which could be used by the Orapa diamond mine, which lies 250 kilometres (150 miles) to the southeast of the delta. The mine's water requirements may double by the year 2004. In the face of these demands for water, the Government explored the possibility of increasing the use of the delta's water. However, extensive analysis has shown that large structural investment in dredging and damming would leave most people worse off. It has further shown that such costly investment may not be necessary. Through more careful management of the existing water supply, and use of groundwater for the town and diamond mine, no dams would be required. This would leave the way open for an integrated regional development programme based upon sustainable use of the Okavango and its resources.

Left **The vast Okavango Delta stretches over some 28,000 sq km (11,000 sq miles). About half the land is permanent swamp, the rest is seasonal floodplain grassland, which is ideal for cattle ranching. Although measures have been taken to preserve some areas in their natural state, signs of change are readily apparent. In this aerial photograph, taken near the town of Maun, Botswana, fenced areas can clearly be seen (unnatural straight lines and right angles) on the left-hand side of the picture.**

Above **Having swooped down from its perch overlooking the open water, an African fish eagle pulls out of its glide and seeks height once more, its prey firmly held in powerful talons. The bird is more often heard than seen, thanks to its far-carrying, gull-like cry. Although fish up to 2 kg (4.5 lb) are the usual prey, the African fish eagle will also feed on rodents, small birds and carrion when its normal prey is hard to find. Harassing prey from other fish-eating birds is also a favoured tactic.**

Northern Asia

Stretching over 6,000 kilometres (4,000 miles) from the Ural Mountains in the west to the Pacific coast in the east, and 5,000 kilometres (3,000 miles) from the Arctic Ocean to the mountains of Afghanistan, northern Asia includes some of the most extensive wetlands in the world. Mirroring the higher latitudes of North America, the tundra of northern Asia presents a landscape of fens and bogs interspersed with lakes, river valleys and coastal estuaries. To the south, peat bogs are the main wetland feature of the predominantly mountainous forest zone. However, extensive productive wetlands exist along river floodplains. In western Siberia, for example, the floodplain of the River Ob extends over 50,000 square kilometres (20,000 square miles). It supports the largest waterfowl breeding and moulting area in Euroasia and is home to the Siberian crane (*Grus leucogeranus*) and white-tailed eagle (*Haliaeetus albicillla*).

Further south, the steppe and desert regions surprisingly support many freshwater lakes with extensive reed stands and a variety of other aquatic plants. These regions also comprise several salt lakes. The largest number of wetlands lies in the south of the Siberian lowland and in northern and central Kazakhastan. Here, flamingo, white-headed duck (*Oxyura leucocephala*), mute swan (*Cygnus olor*), ruddy duck (*Oxyura jamaicensis*), shelduck (*Tadorna tadorna*), great white and Dalmatian pelicans (*Pelecanus onocrotalus* and *P. crispus*) all nest in the wetlands.

In the high mountains of north central Asia, wetland ecosystems are largely confined to oligotrophic lakes (rich in oxygen, poor in nutrients) and fast mountain streams. The birdlife is, however, unusual, with ibisbill (*Ibidorhyncha struthersii*), bar-headed goose (*Anser indicus*) and brown-headed gull (*Larus brunnicephalus*) being typical inhabitants of the region.

At the end of the breeding season there are estimated to be over 40 million waterfowl using the wetlands of northern Asia. The tundras of the northeast are especially important and provide nesting habitats for the emperor goose (*Philacte canagica*), common, spectacled and Steller's eiders (*Somateria mollissima, S. fischeri* and *Polysticta stelleri*) and snow geese (*Anser caerulescens*), while the Taymyr Peninsula provides the breeding grounds for geese such as brent, red-breasted (*Rufibrenta ruficollis*), white-fronted (*Anser albifrons*) and bean geese (*A. fabalis*). Bird diversity is highest in the eastern regions, where European and northern Asian species mix with American species from the northeast, and Chinese and Indo-Malaysian species from the southeast. Baikal and falcated teal (*Anas formosa* and *A. falcata*) and velvet scoter (*Melanitta nigra*) are widespread breeding species, while in the Amur, swan goose, Baer's pochard (*Aythya baeri*), eastern white stork (*Ciconia boyciana*), Manchurian cranes (*Grus japonensis*) and hooded cranes (*G. monacha*) form a rich breeding avifauna.

Today's threats

For centuries, the wetlands of the region avoided any serious human impacts. However, today, oil and gas exploration in western Siberia have resulted in significant pollution and transformation of the landscape. In the east, forestry and mining activities have had significant impacts upon the flora and fauna of the river basins.

1 Lake Baikal

At 1,620 metres (5,315 feet), Lake Baikal is the world's deepest lake. Two-thirds of the species found here are unique, the most famous is the Baikal seal (*Phoca sibirica*).

Map Note

The peatland areas of northern Asia have not been mapped as sufficiently accurate maps are unavailable.

Ramsar Sites
Coastal Wetlands
Water Bodies
Deltas
Freshwater Marsh,
Floodplains

In the steppe region, intensive grazing and expanding agricultural activities in the catchment, diversion of water for irrigation and pollution by fertilizers and pesticides, as well as excessive fishing in many of the lakes, are all widespread problems which need to be addressed. The degradation of the wetlands has been compounded by the drier conditions since the 1850s. In the past, the presence of very wet years with high-water levels was a critical element in preventing siltation and in maintaining the productivity of the lakes. Today, with lower flood levels this process of regeneration has been halted and the productivity of the wetlands is declining.

The Caspian region
The shallow waters of the shoreline of the Caspian Sea support an abundant waterfowl population and a rich fish fauna all year round. The highest natural production of sturgeon, for example, is from the Caspian, which holds 25 per cent of the world's sturgeon species. The Caspian Sea's most important wetland system in the area is the Volga Delta, which in this generally arid region supports more than 250 bird species, 60 species of mammal, including the endemic Caspian seal (*Phoca caspica*), and 80 species of plants. West of the delta, the shallow sea supports extensive underwater meadows of the grass *Chara nitella*, which is the main food plant of the 350,000 or so mute swans that visit the area in summer and autumn. A rich diversity of fish-eating birds depends upon the productive fish resources of the Caspian shores. Approximately 40,000 pairs of great black-headed gulls (*Larus icthyaetus*), 80 per cent of the world's entire population, breed in the Caspian wetlands, together with approximately 6,000 pairs of sandwich terns (*Sterna sandvicensis*) and 250 pairs of Dalmatian pelican.

0 km   1,000   2,000

0 mile   500   1,000

O C E A N

Taymyr Peninsula   *Laptev Sea*

East Siberian Sea

S i b e r i a

Lena   *Indigirka*

Kolyma

R U S S I A N   F E D E R A T I O N

Yenisey   *Bering
Sea*

Magadan

Lena   *Sea of
Okhotsk*

1
*Lake
Baikal*
Irkutsk

*PACIFIC
OCEAN*

CHINA

Ulaanbaatar •   **Zhalong Reserve**

MONGOLIA   **Lake Khanka**   **Kutcharo-ko**

Xiang Hai Reserve   • Harbin   **JAPAN**   **Kushiro-Shitsugen**

Left **A local woman poses with her camels in front of the rusting hulks of Aral Sea ships. Even the most apocalyptic images can do little to convey the enormity of the disaster that has overtaken the Aral Sea. Fishing fleets have literally been replaced by ships-of-the desert. Life-giving water has been replaced by dry sand.**

The Bi-Ob' Area

The lower Ob' valley forms a labyrinth of intricately arranged channels and floodplain lakes, intermingled with meadows and shrubs. The valley stretches 740 kilometres (445 miles) between its junction with the Irtysh River and its estuary at the head of a long fjord stretching to the Kara Sea. Referred to as the Bi-Ob' region, this vast floodplain was periodically inundated by the sea as recently as 20,000 years ago. As the sea gradually receded, the Bi-Ob' developed as a long, stretching pseudodelta, whereby a delta-like feature was formed by the sea receding rather than the river depositing silt. Today, this delta is the largest single breeding area for waterfowl in Eurasia. At the end of the breeding and moulting season some 6 million birds use the area, while several hundred thousand shorebirds, ducks and geese stop here before they continue migrating.

In common with other seasonal floodplains, the Ob' region is a land of fluctuating waterlevels, with both seasonal and annual fluctuations in river discharge and flooding patterns. These flood cycles have a pronounced effect upon the wildlife of the area. In years of moderate flooding, breeding and moulting waterfowl are widely distributed within the Bi-Ob' area, while in years of low floodings they move after nesting to moult in the Ob' estuary. At times of exceptionally high water levels, breeding is confined to higher, peaty areas in the floodplain and to small tributary valleys, while non-breeding birds usually move south into the forested steppe.

Oil exploitation

Like its United States counterpart on the Alaskan North Slope, the Bi-Ob' system is situated at the centre of an oil- and gas-rich province. Great care will be needed if the exploitation of these resources does not lead to pollution of the river system and to a loss of diversity and reduced productivity of the fish and waterfowl populations of the region.

Above **The speed and scale of ecological change in the Aral Sea region is unparalleled in human history. This photograph shows an area of what was once sea bed. The former bottom ooze has desiccated into a fine sand that is already starting to form drifts and dunes and desert plants have replaced the seaweed.**

Below **Bewick swans blend into their Arctic nesting ground in Siberia. In winter, they will migrate to central China, away from the snow. Few of the large waterfowl that nest in northern Asia migrate any great distance. By contrast, the smaller waterbirds that nest on the Arctic tundra, migrate as far as South Africa and New Zealand.**

Aral Sea

Located in the heart of arid Central Asia, the Aral Sea was until recently the fourth largest of the world's lakes. Its shores were broken by the extensive deltas of the Syr and Amu rivers, two of the most biologically diverse wetland systems in the region. Today the Aral is much changed. Its surface area has been reduced from 64,500 square kilometres (25,000 square miles) before the 1950s, to 22,000 square kilometres (8,500 square miles).

The degradation of the Aral Sea is regarded as one of the world's worst ecological catastrophes. For thousands of years the people of Central Asia used the waters of the Syr and Amu rivers to grow crops in desert oases, to which they diverted water through irrigation. These relatively small interventions had little impact on the flow into the Aral. This changed, however, in the 1960s, with the construction of much larger canals designed to irrigate land hundreds of kilometres away. As a result, the areas of irrigated agriculture in Central Asia and Kazakhstan grew from 290,000 square kilometres (110,000 square miles) in 1950 to 720,000 square kilometres (280,000 square miles) by 1992; while the average annual inflow to the sea dropped from over 50 cubic kilometres (11 cubic miles) between 1911 and 1960, to only 5 cubic kilometres (1 cubic mile) from 1981 to 1985. In 1986, a relatively dry year, no water reached the Aral Sea at all. Between 1960 and 1989, the level of the sea fell by 14 metres (46 feet), its area by 45 per cent, its volume by 68 per cent and the salinity of its waters increased from 10 to 28 grams per litre (1.6 to 4.5 ounces per gallon). By 1990 the Aral had divided into two parts.

Far-reaching impacts

The biological, social, and economic impact of these changes to the Aral Sea region have been disastrous. The once valuable annual fish catch has disappeared altogether, while the harvest of over a million muskrat pelts has also fallen to virtually zero. In addition, the effects of the shrinkage to the Aral Sea are not just localized. Large water bodies have a moderating effect on the neighbouring climate, and today, with the decline of the sea, the climate of the region is becoming increasingly severe. It has been estimated that the growing season has declined by 10 days in the northern reaches of the Amu Darya Basin, some 200 kilometres (120 miles) away.

In the face of this array of problems there is widespread national and international concern for the future of the Aral Sea and the people of the region. In October 1990, an international symposium was held in Nukus in the Karakalpak Autonomous Soviet Socialist Republic to discuss the causes of the crisis, its impacts, and possible solutions. In their concluding resolutions, scientists argued that ecological restoration of the sea is impossible unless the hydrological cycle of the area is stabilized and reverted to something like its natural state. This could only be achieved by severely restricting the amount of water diverted from the rivers for irrigation. However, these measures will need to be accompanied by major institutional reforms which provide farmers of the region with more latitude in selecting the crops they grow and the farming practices they employ, as well as to provide greater incentives for farmers to conserve water and to reduce chemical inputs in crop production.

Central and South Asia

The Himalayan mountain range dominates southern Asia. Its snowfields and glaciers give rise to some of the world's mightiest rivers, among them the Ganges, Brahmaputra, Indus, Mekong and Yangtze. For centuries, these massive freshwater arteries have nourished some of the most densely populated areas on Earth. The silt enriched valleys and huge deltas of the Indus and Ganges/Brahmaputra, for example, contain extensive wetlands that have sustained sophisticated civilizations for around 4,500 years.

Today, although modern management, technologies and practices have altered river flow and destroyed some of the different types of wetland, they continue to play a critical role in supporting the people of the continent. In Bangladesh, for example, more than 5 million people are dependent on fishing for their livelihood. The annual harvest of fish, crustaceans and frogs is estimated as being between 675,000 and 725,000 tonnes, of which 81 per cent comes from rivers and wetlands, while the remainder is derived from marine fisheries. In India, the fishery of Chilka Lake in Orissa State alone yields 700 tonnes of fish per year, and provides the principal livelihood for people living along its shores.

High in the Himalayas, the main wetlands are the mountain lakes, some of which have fringing marshes. Although most of these lakes are poor in nutrients and support only a limited flora and fauna, they are of great cultural importance and are held sacred by the Hindu and Buddhist religions.

In Afghanistan, the combination of mountainous relief over much of the country and the arid climate of the southwest have confined wetlands to the major rivers which rise in the mountain ranges of the north. The largest single wetland is the Hamoun-e Puzak, one of a group of three large freshwater lakes in the Seistan Basin – an inland drainage basin surrounded by desert and lying on the border with Iran. These wetlands receive most of their water from the Helmand River, which rises far away to the northeast in the Hindu Kush. The only other large wetlands are two salt lakes in the eastern highlands, Dashte Nawar and Ab-i Estada, both of which are renowned for their flamingos.

By 1979, four reserves with important wetlands – Ab-i Estada Waterfowl Sanctuary, Dashte Nawar Waterfowl Sanctuary, Bande Amir National Park and Kole Hashmat Khan Waterfowl Sanctuary – had been established. However, conservation activities were brought to an abrupt halt in 1979 due to political unrest, and it is doubtful if any practical conservation measures have been implemented since then.

Further south, in Pakistan, the main wetlands lie along the Indus Valley. These range from fresh to slightly brackish lakes and ponds (many of which provide water for urban consumption), saline and freshwater marshes, which have been formed by water from irrigation canals, to the estuary and delta at the mouth of the Indus. These wetlands are of critical importance across the country, but the pressure for land and water along the Indus is placing them under increasing pressure.

India's wetlands

India is the giant of the sub-continent and with a population of over 800 million people is the world's second most populous nation. Most of the people are farmers and are concentrated in the fertile plains and valleys which support the country's principal wetlands. These include the reservoirs of the Deccan Plateau in the south, together with the lagoons and other wetlands of the southern west coast; the vast saline expanses of Rajasthan, Gujarat and the Gulf of Kutch; freshwater lakes and reservoirs from Gujarat eastwards through Rajasthan and Madhya Pradesh; the delta wetlands and lagoons of India's east coast; the freshwater marshes of the Gangetic Plain; the floodplain of the Brahmaputra and the marshes and swamps in the hills of northeast India and the Himalayan foothills; the lakes and rivers of the montane region of Kashmir

◼ The Mahaweli Ganga

The most extensive floodplain system in Sri Lanka is the marsh area of the Mahaweli Ganga, the largest river in Sri Lanka. The region comprises small, individual wetlands known as "villus". Twenty-six of these villus, the largest of which is 8 sq km (3 sq miles), lie within the boundaries of national parks, and constitute a linked system of protected areas. The floodplains provide a migratory corridor between wet and dry season feeding grounds for the largest concentration of elephants (*Elephus maximus*) and they are home also to the highest density of large mammals in the country, including the endangered leopard (*Panthera pardus*), toque macaque (*Macaca sinica*) and sloth bear (*Melursus ursinus*). Reptiles include the threatened python (*Python molurus*), estuarine crocodile (*Crocodylus porosus*), marsh crocodile (*C. palustris*) and the endemic lizards (*Calotes zeylonensis* and *Otocryptis weigamanni*), while amphibians include the palm-frond frog (*Hylerana gracilis*). Most of the 250 resident species of birds have been recorded from this floodplain system, as well as 75 species of migratory birds. Dams constructed on the Mahaweli have altered the flow to several of the villus, but despite initial concerns they have retained their biological importance.

FORMER

AFGHANISTAN

Bande Amir National Park

Kole Hashmat Khan Waterfowl Sanctuary

Dashte Nawar Waterfowl Sanctuary

Ab-i Estada Waterfowl Sanctuary

Kabul

PAKISTAN

Drigh Lake

Kinjhar (Kalri) Lake

Kot

Haleji Lake

Arabian Sea

Rat

Gulf of Kutch

◼ Ramsar Sites
◻ Parks and Reserves
◻ Water Bodies
◻ Mangroves
◻ Estuaries, Coastal Wetlands
◻ Pans, Brackish Wetlands
◻ Swamp Forest
◻ Tank Regions
◻ Wetland Complexes
◼ Freshwater Marsh, Floodplains

0 km	500	1,000

0 mile	250	500

INDIAN

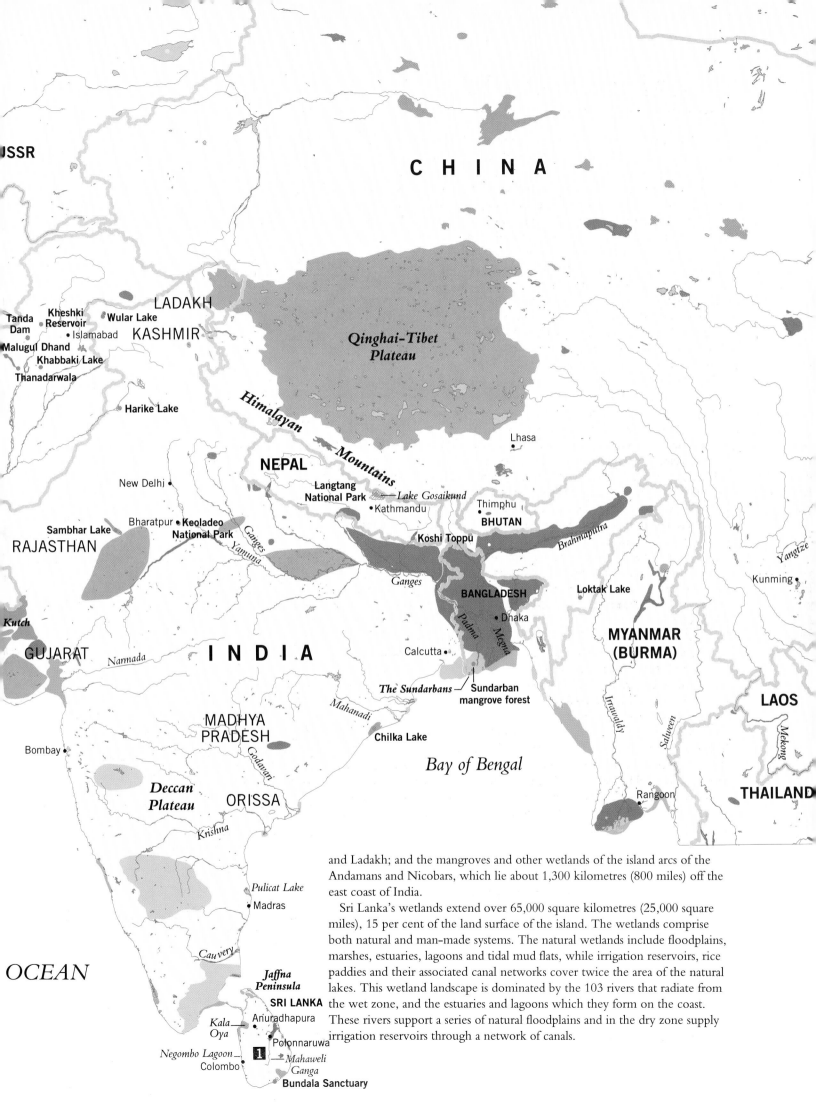

USSR

CHINA

LADAKH
KASHMIR

Tanda
Dam
Kheshki
Reservoir
• Islamabad
Wular Lake

Malugul Dhand
Khabbaki Lake

Thanadarwala

Harike Lake

*Qinghai-Tibet
Plateau*

Himalayan

Mountains

Lhasa

NEPAL

New Delhi •

Langtang
National Park
~ *Lake Gosaikund*
• Kathmandu

Thimphu •

BHUTAN

Sambhar Lake

Bharatpur • Keoladeo
National Park

Ganges

Yamuna

Koshi Toppu

Brahmaputra

Yangtze

RAJASTHAN

Ganges

BANGLADESH

Kunming •

Loktak Lake

Kutch

GUJARAT

Narmada

INDIA

Padma

• Dhaka

Megna

MYANMAR
(BURMA)

Calcutta •

MADHYA
PRADESH

Mahanadi

The Sundarbans

Sundarban
mangrove forest

LAOS

Bombay •

Godavari

*Deccan
Plateau*

ORISSA

Chilka Lake

Bay of Bengal

Irrawaddy

Salween

Mekong

THAILAND

Krishna

Rangoon •

Cauvery

Pulicat Lake

• Madras

OCEAN

*Jaffna
Peninsula*

SRI LANKA

*Kala
Oya*

Anuradhapura

Negombo Lagoon
Colombo

Polonnaruwa

1

← *Mahaweli
Ganga*

Bundala Sanctuary

and Ladakh; and the mangroves and other wetlands of the island arcs of the
Andamans and Nicobars, which lie about 1,300 kilometres (800 miles) off the
east coast of India.

Sri Lanka's wetlands extend over 65,000 square kilometres (25,000 square
miles), 15 per cent of the land surface of the island. The wetlands comprise
both natural and man-made systems. The natural wetlands include floodplains,
marshes, estuaries, lagoons and tidal mud flats, while irrigation reservoirs, rice
paddies and their associated canal networks cover twice the area of the natural
lakes. This wetland landscape is dominated by the 103 rivers that radiate from
the wet zone, and the estuaries and lagoons which they form on the coast.
These rivers support a series of natural floodplains and in the dry zone supply
irrigation reservoirs through a network of canals.

Diverting the Indus

For the last 6,000 years, from the Indus civilizations of Moenjodaro to the present day, the River Indus has been the lifeblood of the arid region that is now Pakistan. However, while earlier peoples used the river's waters to cultivate the natural floodplains of the Indus, the past 100 years has seen the Indus progressively dammed and its waters diverted into one of the largest and most complex irrigation systems in the world. By 1992, there were three storage reservoirs, 16 dams, more than 40 canals totalling 56,000 kilometres (35,000 miles), and more than 1.5 million kilometres (900,000 miles) of farm channels and water courses.

With these investments the Indus has become the "bread-basket" for most of modern Pakistan's 100 million or so people. However, the future of this productivity is threatened. In the absence of a drainage system to remove irrigation water, exacerbated by leakage of water from the banks of unlined canals, the water table has been rising in many parts of the valley. When the water table reaches the surface, it evaporates, leaving the salts carried with it in the top soil. This salinization of the soil has, together with waterlogging, resulted in the loss of 400 square kilometres (150 square miles) of irrigated land each year and a total of 57,000 square kilometres (22,000 square miles) are now affected by salinity.

Land loss

The diversion of water is also threatening the future of the 6,000 square kilometres (2,300 square miles) of the Indus Delta. Presently 72 per cent of the Indus' water is withdrawn for irrigation, leaving only 28 per cent to be discharged below the Kotri, the lowest barrage on the river. Because most of this flow to the delta occurs during the monsoon (June to September), the Indus does not flow out into the sea for the rest of the year. The dams also retain the silt carried by the river and only 100 million tonnes (25 per cent) now reaches the delta. As a result, the front edge of the delta is beginning to erode, and future sea-level rise will only exacerbate the gradual eating away of the delta.

Combined with high evaporation, the reduced freshwater flow has raised salinity in many of the delta's creeks to 40–45 parts/1,000, which is higher than seawater (35 parts/1,000), while soil salinity is as high as 70 parts/1,000. As a result, the future of the delta's 2,600 square kilometres (1,000 square miles) of mangrove forest looks uncertain. The high salinity stunts growth and kills seedlings. In turn the biological productivity of the delta, and in particular the fish and crustaceans which use the mangroves as a nursery area, are in danger of dying out.

Above **Fishing boats fly their colourful flags on the Indus Delta in southern Pakistan. Consideration of the Indus and its vast water resources tends to focus on its provision of irrigation water, essential** for the well-being of millions of people. However, the continued flow of fresh water into the Indus Delta is vital if the productivity of the delta's valuable fisheries is to be maintained.

Below **A Bengal tiger prowls across swampy ground in northeast India. Although tigers are conventionally portrayed as rain forest animals, they generally** prefer grassland or even swampland habitats. Tigers, surprisingly, are very capable swimmers, and will readily take to the water should the chase of the prey require it.

The Sundarbans

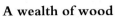

In a land where three of Asia's mightiest rivers, the Ganges, Brahmaputra and Meghna, mingle before flowing into the Bay of Bengal, the Sundarbans constitute the single most extensive mangrove forest in the world. Straddling the border between India and Bangladesh, the forest covers around 6,000 square kilometres (2,300 square miles). The forest floor is threaded with a complex network of rivers, creeks and canals which flood twice daily as the tide rises, creating a rich habitat for the many species of fish and invertebrates that move into the forest with the tides.

A wealth of wood

Twenty-seven species of mangrove tree grow in this dynamic system, with *Heritiera fomis* and *Excoecaria agallocha* in single or mixed stands covering more than 70 per cent of the forest. In a region where the remaining areas of forest are under pressure from the rapidly growing human population, these mangroves make up 45 per cent of the country's productive natural forest and are by far the single largest source of both wood and other forest products. Each year, the forest yields 68,000 cubic metres (2.4 million cubic feet) of timber, 183,000 cubic metres (6.4 million cubic feet) of pulpwood and 106,400 tonnes of fuelwood. There is a rich fish catch all year round, averaging about 150,000 tonnes, and during the winter months about 10,000 fishermen gather at the island of Dubla for fishing within the forest area and adjacent coastal waters. Other products include about 3,000 tonnes of shells which are converted into lime, and about 200 tonnes of honey and 50 tonnes of beeswax which are harvested each year. Over 300,000 people use the forest each year in pursuit of these diverse resources with many others employed in the wood processing industry.

Despite this heavy use of the forest, it continues to support a diverse fauna of 35 species of reptile, over 270 birds and 42 mammals. Most famous among the mammals is the Royal Bengal tiger (*Panthera tigris*) for which the Sundarbans is the last remaining stronghold. It is estimated that between 350 and 450 tigers are still present in the Bangladesh portion of the forest, with a further 250 to 300 in India.

Left **Moored in one of the many creeks of the mangrove forest, a party of tourists peer expectantly into the Sundarbans Tiger Reserve, India. As well as a few hundred tigers, the reserve also contains sizeable populations of other rare animals, including rhesus macaques (*Macaca mulatta*) and spotted deer (*Axis axis*).**

Wetlands and Floods in Bangladesh

About 80 per cent of Bangladesh, an area of some 115,000 square kilometres (44,000 square miles), is formed by the floodplains of the Ganges, Brahmaputra and Meghna rivers. In an average year, 26,000 square kilometres (10,000 square miles) of the floodplain are submerged, while in the largest floods of recent times this rose to 82,000 square kilometres (32,000 square miles) – 57 per cent of the whole country. In addition to this massive floodplain, there are about 700 rivers and lakes which are known locally as *haores* and *beels*, and 6,100 square kilometres (2,400 square miles) of estuarine wetlands, with the Sundarbans mangrove forest forming the single most important system.

Bangladesh's wetlands play a major role in the lives of the people and economy of this very densely populated country. Availability of water for irrigation in the dry season makes it possible to grow up to three crops a year in certain areas. This enables Bangladesh's relatively small agricultural land area to produce most of the basic food requirements for the country's 110 million people. In addition, the deposition of water-borne sediments keeps the soil fertile and in some cases can even enrich it. Crop production, therefore, remains at a satisfactory level without the need for potentially harmful fertilizers. Flooding also promotes algal growth which transfers up to 30 kilograms of nitrogen per hectare (27 pounds per acre) per year from the atmosphere to the soil. In addition, fish provide somewhere in the region of 80 per cent of the daily animal protein intake of the people, while the fishery sector's contribution to the country's Gross Domestic Product (GDP) and export earnings is 6 and 12 per cent respectively. The sector also generates full-time employment for about 2 million people and part-time employment for another 10 million.

Flood control

Despite the importance of these extensive wetlands, and the close dependence upon them of the people of Bangladesh, the country is better known outside of the Asian sub-continent for the damage inflicted upon it by floods and cyclones. Since Bangladesh was established in 1971, a series of massive floods and devastating cyclones have killed more than 500,000 people and left millions homeless.

In response the international community is currently investing about US$150 million to prepare proposals for investment to control the floods and protect people from cyclones. Yet, with 55 per cent of the land surface lying less than 3 metres (10 feet) above sea level at the mouth of three of the world's largest rivers, Bangladesh is a land of floods. The people of the delta have adjusted to this age old phenomenon, and live with the rhythm of the flood. Indeed, in a country of 110 million people where fertile, agricultural land is at a premium, farmers await the flood with anticipation because it brings nutrients and water for agriculture. However, the close relationship between the farmers and the flood is becoming strained. Evidence gathered over the last 40 years suggests that extreme floods are becoming increasingly frequent. Since 1954, severe floods have submerged an area of between 35,000 square kilometres (13,500 square miles) and 82,000 square kilometres (32,000 square miles) in 13 years. In 1988, when 82,000 square kilometres (32,000 square miles), 57 per cent of the country, was flooded, over 2,000 lives were lost and damage to property and infrastructure was estimated at US$1.1 billion. The dilemma facing the Bangladesh Government is how best to invest in reducing the effects of these events.

Working with water

Current proposals focus upon the construction of dikes and other structures which would close off valuable land and habitation from the flood. However, this approach has been the subject of strong criticism from a wide audience, including some highly respected international authorities on water

Above **Fishing a pond near Bogra, Bangladesh. Few fish are likely to be present, and frogs will form the bulk of the catch. Animal protein from wetlands, like this pond, provide a significant proportion of the rural diet throughout South Asia.**

Below **Floodplain under cultivation near Dacca. For centuries, farmers relied on the annual floods. The floods are still important, and the challenge for the future is to ensure that development works within the constraints of this natural force.**

Below **Clearing water hyacinth (*Eichornia crassipes*) by hand in Bangladesh. Introduced from South America, the water hyacinth has become an aquatic pest over much of Asia. Although cutting will create a temporary patch of clear water, the only effective long-term solution is to dig up the thick fibrous root of each plant individually.**

management. Some argue that, rather than invest huge sums in controlling nature, it would be best to work with nature. They point to the fact that in 1757 the region already had a system of small dikes which prevented major damage. Then people simply took to boats if the waters got too high, while benefiting from the fertilizing role of the flood when the water receded.

The issues are certainly complex. The Government of Bangladesh faces the dilemma of needing to protect its population from flood and minimizing damage to infrastructure, while at the same time ensuring that the benefits of the flood are not altogether lost. However, the issue of debate is whether this is indeed possible in a river system, the combined flow of which is second only to the Amazon, and the position of which has been known to move several kilometres in a season. In addition, some authorities estimate that while the urban rich will benefit, over 6 million of the country's poorest people would suffer.

East Asia

With a population of 1.3 billion projected to rise to 2 billion by the year 2025, East Asia currently accounts for approximately 24 per cent of world population. For centuries, the high productivity of East Asia's wetlands, which today still cover some 280,000 square kilometres (108,000 square miles), has played a central role in sustaining the living standards of hundreds of millions of people. For example, almost 95 per cent of China's population of 1.1 billion is concentrated in the eastern half of the country, principally in the vast alluvial plains of the major rivers.

China has more than 250,000 square kilometres (97,000 square miles) of wetlands. These include marshes and bogs, lakes (both natural and artificial), and coastal salt marshes and mud flats. This diversity can be grouped into five principal systems: the high-altitude lakes and bogs of the Qinghai-Tibetan Plateau, the inland drainage systems of the Sinkiang Basin, the freshwater lakes, marshes and peat bogs of Heilongjiang, Jilin, Lianing and Inner Mongolia in the northeast; the lake systems and riverine wetlands along the middle and lower reaches of the Yellow and Yangtze rivers and finally the estuarine, mud flat and mangrove systems along the coast.

China's 21,000 square kilometres (8,000 square miles) of coastal marshes and mud flats are concentrated in three main areas: at the mouth of the Yangzte River and along the adjacent coast of Jiangsu Province, around the estuary of the Yellow River in Bohai Gulf and in the estuarine system of the Shuangtaizi and Liao rivers in Liaoning Province. Only a small proportion of what are thought to have been extensive wetlands now remain. Scant stands of mangroves grow along the coast as far north as central Fujian Province and also on the island of Taiwan, but many of these mangrove/mud flat ecosystems have now been converted to rice paddies and aquaculture ponds. One of the richest surviving stands of mangrove is in Deep Bay, on the border between Guangdong Province and Hong Kong.

Diverse wetlands

Despite rainfall that averages only 100–400 millimetres (4–16 inches) a year, Mongolia is rich in water resources mainly because of the high mountain ranges which attract precipitation. There are approximately 15,000 square kilometres (6,000 square miles) of standing water bodies and 50,000 kilometres (30,000 miles) of rivers. Wetland habitats are extremely diverse, ranging from cold, deep ultra-oligotrophic (poor in nutrients and rich in oxygen) lakes to temporary saline lakes, and there are many major rivers possessing extensive floodplains. Only the southern desert margin and the southeast of the country lack any permanent water. Many of the lakes and marshes are extremely important breeding and staging areas for migratory waterfowl, notably ducks, geese and cranes. Some of these birds enjoy special protection that dates back to the 1200s when laws were enacted forbidding the hunting of game during the breeding season. In addition, human population pressure remains low throughout much of Mongolia and most wetlands are still in an almost pristine condition, disturbed only by the occasional hunter, fisherman or shepherd.

1 Mongolia's Lakes

Most of Mongolia's principal lakes are situated in the Central Asian Internal Drainage Basin. Some 144 million years ago, this vast basin formed an inland sea which covered the entire region. As the humid climate of these prehistoric times gave way to more arid conditions, however, the sea gradually broke up to form the relict lakes that remain today. These include freshwater lakes with outlets into others such as Ayrag and Har-Us lakes, and saline lakes such as Hyargan and Dorgon which have no outlet. The fish fauna of the basin reflects its history. Many are relict species that are left over from the ancient sea.

FEDERATION

Sea of
Okhotsk

Amur

Amur

Amur

Sakhalin

HEILONGJIANG

**Sanjiang
(Three Rivers)
Plain**

Hulun Lake

*Buyr
Lake*

Zhalong Reserve **Zhalong Marshes**

Sungari

Wusuli
Ussuri

Kutcharo-ko
Hokkaido

Kushiro-
shitsugen

Xiang Hai Reserve

*Xiang Hai
Marshes*

JILIN

*Lake
Khanka*

Khanka Reserve

Utonai-ko
•Kushiro

• Jilin

Liao

Taizi

LIAONING

**NORTH
KOREA**

Sea of
Japan

ONGOLIA

SHANXI

•Beijing

Chongchon-Gang
Taedong

JAPAN

Izu-numa and
Uchi-numa

I N A

HEBEI

Korea Bay

•P'Yongyang

Honshu

*PACIFIC
OCEAN*

HAANXI

Bohai Gulf

Kaesong
Seoul

Tokyo•

Bay of Tokyo

SHANDONG

Imjin Basin
Han Basin

•Nagoya

HENAN

Han

Nakdong

**SOUTH
KOREA**

West Taegu

Lake Biwa

Inner Ise Bay
Ise Bay

Huang (Yellow)

Yashiro

*Osaka
Bay*

*Yellow
Sea*

JIANGSU

ANHUI

East
China
Sea

Izumi •

• Shanghai

HUBEI

*Wuhan
Lakes*

Yangtze

*Shengjin
Lake*

Kyushu

Yangtze

**Dongtinghu
Nature Reserve**

Poyang Lake Reserve

Dongting Lakes

Poyang Lake

ZHEJIANG

HUNAN

JIANGXI

FUJIAN

TAIWAN

	0 km	500	1,000
	0 mile	250	500

The wetlands of Korea and Japan

In the Korean Peninsula the principal wetlands are along the west and south coasts. Here there are numerous estuaries and shallow sea bays with extensive intertidal mud flats and offshore islands. In some areas there is a broad coastal plain with many small lakes, expansive reedbeds and large areas of rice paddy. This marshy coastal plain is especially well developed around the estuaries of the Chongchon-Gang and Teadong rivers in the north, and in the lower basins of the Imjin, Han and Nakdong rivers in the south. These wetland systems are now under heavy pressure. Of the 487 natural lakes and ponds in Japan, most are very small. However, the largest lake in Japan, Lake Biwa in central Honshu, is the notable exception. This particular lake covers 674 square kilometres (260 square miles). Eastern Hokkaido supports the most extensive freshwater marshes remaining in Japan, as well as the majority of the country's remaining lagoons and salt marshes.

Elsewhere, most of the lowland marshy habitats and coastal lagoons have been drained for agriculture. The tidal ranges of Japan's Pacific coast sustain the country's largest area of intertidal mud flats. Estuaries and bays such as Tokyo Bay and Inner Ise Bay also once supported extensive mud flats, but much of this habitat has been lost to urban development. Mangrove swamps are confined to the Amami Islands and Ryukyu Islands, where they grow on muddy beaches and in estuaries.

GUANGDONG

Guangzhou

**Mai Po Marshes
Nature Reserve**

Xi Jiang

HONG KONG

Macao •

• Victoria

	Ramsar Sites
	Parks and Reserves
	Water Bodies
	Mangroves
	Estuaries, Coastal Wetlands
	Pans, Brackish Wetlands
	Swamp Forest
	Wetland Complexes
	Freshwater Marsh, Floodplains

**Dongzhaigang
National
Nature
Reserve**

Hainan

South
China

Sea

Lakes and the Three Gorges Dam

The greatest concentration of freshwater lakes in China occurs on the alluvial plains along the middle and lower reaches of the Yellow and Yangtze rivers. The total area of lakes is estimated at over 22,000 square kilometres (8,500 square miles). The most important include the Dongting Lakes in Hunan Province, the Wuhan Lakes in Hubei Province, Poyang Lake in Jiangxi Province and Shegjin Lake in Anhui Province, all in the Yangtze Basin. The lakes are used intensively by local people for fishing, grazing and farming, practices that do not threaten the internationally important populations of waterbirds, such as the Siberian crane (*Grus leucogeranus*), which utilize the lake in winter. The world population of the Siberian crane was thought to number only a few hundred until a wintering flock was discovered at Poyang Lake in 1980. Numbers are now estimated at 2,600 individuals, approximately 95 per cent of the known world population of this species.

The principal threat to these lakes is the alteration of the hydrological regime, either through retention of water in dams upstream or through modification of the flow patterns. One such project is the enormous Three Gorges Dam planned for the Yangtze River. Estimated in early 1989 to cost US$10 billion, the project is expected to generate almost a fifth of China's entire energy needs. Its supporters claim that the dam will control the massive flood that devastates the floodplain of the Yangtze about once every 10 years, while also improving navigation along the Yangtze, China's most important waterway. Environmentalists, however, warn that the project will lead to flooding on a huge scale of the area behind the dam, which in turn will lead to the displacement of more than a million people. The dam is also likely to reduce the flooding of the natural wetlands along the Yangtze Valley, including Poyang Lake.

East Asia's Wildlife

Extending almost from the subarctic to the tropics, the wetlands of China, Mongolia, Korea and Japan support an especially rich and diverse fauna and flora. Over 170 species of waterbird occur in the region, including 36 species of duck and goose, eight species of crane and 53 species of shorebird. Wetlands of Mongolia, the Qinghai-Tibetan Plateau, northeastern China and Hokkaido support large breeding populations of waterbirds, while the lakes, marshes and coastal wetlands of central and southern China, Korea and southern Japan provide important staging and wintering areas for these and many other migratory species of waterbird breeding in arctic Russia.

However, populations of most species of waterbirds and other wildlife have been reduced to a fraction of their former levels. The reductions are a result of the massive loss of wetlands to agriculture and urban development, very heavy hunting pressure and general disturbance from the huge human population. No less than 27 species of waterbird are now listed in the IUCN Red Data Book as threatened, and several of these are nearly extinct.

Asia's crocodile

East Asia's only crocodilian, the Chinese crocodile (*Alligator sinensis*), is on the brink of extinction. The total population of this species was thought not to exceed 2,000 individuals in 1984, their last stronghold being in the lakes and rivers of the lower Yangtze Valley in Anhui Province. The Chinese river dolphin or Baiji (*Lipotes vexillifer*), also confined to the Yangtze River, is in an even worse plight, with a total population possibly numbering less than 100 individuals. Both species have been reduced to these low levels by a combination of direct persecution, and dam and barrage projects on the Yangtze and its tributaries. The Chinese water deer (*Hydropotes inermis*), however, remains fairly common around lakes in the Yangtze Valley and in the extensive coastal marshes of Jiangsu Province.

Above **A mixture of traditional and modern river craft crowd the Yangtze Gorges, near the proposed site of a massive hydroelectricity/flood-control scheme. Although the project will provide about 20 per cent of the nation's energy requirement, and improve navigation, these advantages must be balanced against the environmental effect on downstream wetlands.**

Above right **The oriental fire-bellied toad (*Bombina orientalis*) is found in many East Asian wetlands. Its bright red markings warn potential predators of the irritating milky secretion the toad exudes when threatened.**

Right **Traditional fishing baskets line the shore of Lake Ermai, close to China's southern border. Although the culinary adventurism of the Chinese is well known, with few animal species considered inedible, overfishing of lakes is not generally a problem. Informal management techniques, which have developed over thousands of years, ensure that the lakes provide a sustainable harvest.**

China's Reeds

The wetlands of temperate northeastern China include some of the most extensive reed marshes in Asia. The Sanjiang (Three Rivers) Plain near the confluence of the Heilong (Amur), Sungari and Wusuli (Ussuri) rivers, is the largest single wetland area, with over 10,000 square kilometres (4,000 square miles) of shallow lakes, reed beds and peat bogs. But there are several other very extensive systems of freshwater lakes and marshes such as Zhalong Marshes near Qiqihar and Xiang Hai Marshes in western Jilin. These marshes are critically important habitats for both migratory and breeding cranes, and other waterbirds.

In several areas, reeds are harvested on a commercial basis for the production of high-quality paper. At Zhalong Marshes in Heilongjiang Province, one state-run company employing 450 full-time workers and 1,000 temporary workers harvested 26,000 tonnes of reeds in 1984. Most of these wetlands have remained largely unaltered by human activity. However, proposals to divert the rivers that flood the Sanjiang Plain may lead to the loss of this large floodplain system within the next few years.

Coastal Reclamation in Korea

A severe shortage of land suitable for agriculture in the Republic of Korea is now putting a tremendous strain on the country's coastal wetland regions. Much of the traditional agricultural land has been lost to the rapid development of urban and industrial areas in recent years. Domestic food production has been falling while at the same time the human population has been rising steadily for the last 40 years or so. As a consequence, there has been a steep rise in food imports.

In an effort to solve these problems, the Korean Government is pursuing a major programme of land reclamation at estuaries and shallow bays on the south and west coasts of the country. In a feasibility study carried out by the Government in 1984, 155 estuaries and bays with a total area of some 4,180 square kilometres (1,600 square miles) were identified as being suitable for land reclamation projects. It is anticipated that all 155 sites will eventually be reclaimed, resulting in the loss of approximately 65 per cent of the total coastal wetlands of the country.

The Cranes of East Asia

Eight of the world's 15 species of crane live in East Asia: the common crane (*Grus grus*), black-necked crane (*G. nigricollis*), hooded crane (*G. monacha*), red-crowned crane (*G. japonensis*), white-naped crane (*G. vipio*), sarus crane (*G. antigone*), Siberian crane (*G. leucogeranus*) and demoiselle crane (*Anthropoides virgo*). The sarus crane, primarily a species of southern Asia, is an extremely rare visitor to wetlands in Yunnan Province in southern China, while the common and demoiselle cranes are relatively prevalent and widespread across much of Eurasia, spending winter in Africa and the Indian sub-continent. The other five species are, however, more or less confined to eastern Asia.

The black-necked crane breeds in marshes and peat bogs at elevations of over 3,300 metres (11,000 feet) above sea level on the great Qinghai-Tibetan Plateau in southwest China and in neighbouring Ladakh in India. In autumn, the cranes undertake relatively short migrations to spend the winter months in valleys and marshes at lower elevations, generally between 2,000 and 2,500 metres (6,500 and 8,000 feet) above sea level. Key wintering areas include the Lhasa Valley in Tibet, the Popshika, Boomthang and Tashi Yangtsi valleys in Bhutan, Lu Guhu, Bitahai and Napahai marshes in Yunnan Province, and Caohai (which literally translates as "sea of grass") in neighbouring Guizhou Province. The total population of this rare crane has recently been estimated at about 3,000 individuals.

Migrating cranes

The other four species of crane are northern breeders, undertaking long migrations via a series of traditional staging areas to wintering grounds in eastern China, the Korean Peninsula and southern Japan. The Siberian crane undertakes the longest migration, travelling from breeding areas in the Arctic tundra of Yakutia to its wintering grounds in the Yangtze Valley. The great bulk of the 3,000 or so Siberian cranes spends the winter at Poyang Lake, although small flocks also settle on Shengjin Lake and the Dongting Lakes. Traditional staging areas include Zhalong and Momoge marshes in Heilongjiang Province in northeastern China. Two much smaller populations of Siberian cranes breed further west in arctic Russia and winter in northern India and Iran. Unfortunately, however, both these populations now number no more than about 15 individuals.

The hooded crane breeds in the marshes of the coniferous-forested taiga zone in eastern Siberia and Amurland. While a small proportion of the population migrates through eastern China to winter at Shengjin, Poyang and the Dongting lakes in the Yangtze Valley, the bulk of the population takes a more easterly route through the Korean Peninsula to spend the winter in rice fields at West Taegu in South Korea and at Izumi (Kyushu) and Yashiro (Honshu) in southern Japan. By far the most important site for this species is the famous crane sanctuary at Izumi, where as many as 5,500 of the total population of about 6,500 spend the winter.

The white-naped crane breeds in the extensive wetlands of the temperate steppe zone. This area stretches from eastern Mongolia through Heilongjiang and Jilin Provinces of northeastern China to the Ussuri River along the Russian border. About two-thirds of the population, which numbers some 3,500 birds, take a westerly route to winter in the Yangtze Valley (Shengjin, Poyang and Dongting), while the remainder take an easterly route to winter in the Korean Peninsula (principally at Taesong'dong and Panmunch'om) and at Izumi in Japan.

The red-crowned crane is the most southerly breeder of these four cranes, breeding in reed marshes in Hokkaido, Amurland and northeastern China as far south as the estuarine marshes of the Shuangtaizi and Liao rivers in Liaoning Province. The Hokkaido population, which numbers about 350 to 400 birds, is largely sedentary, the birds moving only a short distance from their breeding marshes to spend the winter around feeding sites near Kushiro.

Below The sarus crane is one of the largest members of the crane family, and mature adults stand more than 1.5 m (5 ft) tall. The sarus crane is distributed over a 1,000-km (600-mile) wide band that stretches from the Hindu Kush to Vietnam. Among the human population of this region, the sarus crane has acquired a protected, almost sacred status. The reason for the veneration of the sarus crane is the close, lifelong pair-bonding that characterizes the species. This bonding is so close that adult cranes are rarely seen alone. To the human population, fond of animal archetypes, the sarus crane has become symbolic of fidelity, a virtue that is as highly regarded in the East as it is in the West.

A population of about 400 birds breeding along the Ussuri River and in Amurland winters primarily in the coastal marshes of South Hwanghae and Kaesong in the Democratic People's Republic of Korea. Western breeding birds, numbering about 650, migrate round the shores of the Bohai Gulf to winter in the vast coastal marshes of Jiangsu Province and to the islands at the mouth of the Yangtze River.

Cranes and golf

Known as *tancho,* the red-crowned crane has become a symbol of conservation in Japan, expanding their traditional role as symbols of longevity and happiness. In Kushiro, one of the major threats to the cranes that breed and winter there comes from an unexpected quarter, golf courses. In the area surrounding Kushiro Marsh, there are seven golf courses and an additional 12 are under construction. Presently, the courses cover an area of some 50 square kilometres (19 square miles). In the interests of high-quality fairways and greens, the courses have been sprayed with pesticides, herbicides and fertilizers, many of which eventually find their way into the marsh downstream. While the precise effects of these chemicals are unknown, the toxic effects of pesticides in many other similar situations have given rise to growing concern for the cranes and other animal species that depend upon the marsh. Conservationists argue that it would be a national tragedy if the *tancho* was to be endangered because of golf.

Above **The demoiselle crane often breeds in marshy riverside areas. Unlike the common crane, which builds an elaborate nest, demoiselles make a scrape in the ground.**

Below **A pair of black-necked cranes on the Caohai Nature Reserve, China. Having no local special status, the eggs of this rare species have long been collected for food.**

Southeast Asia

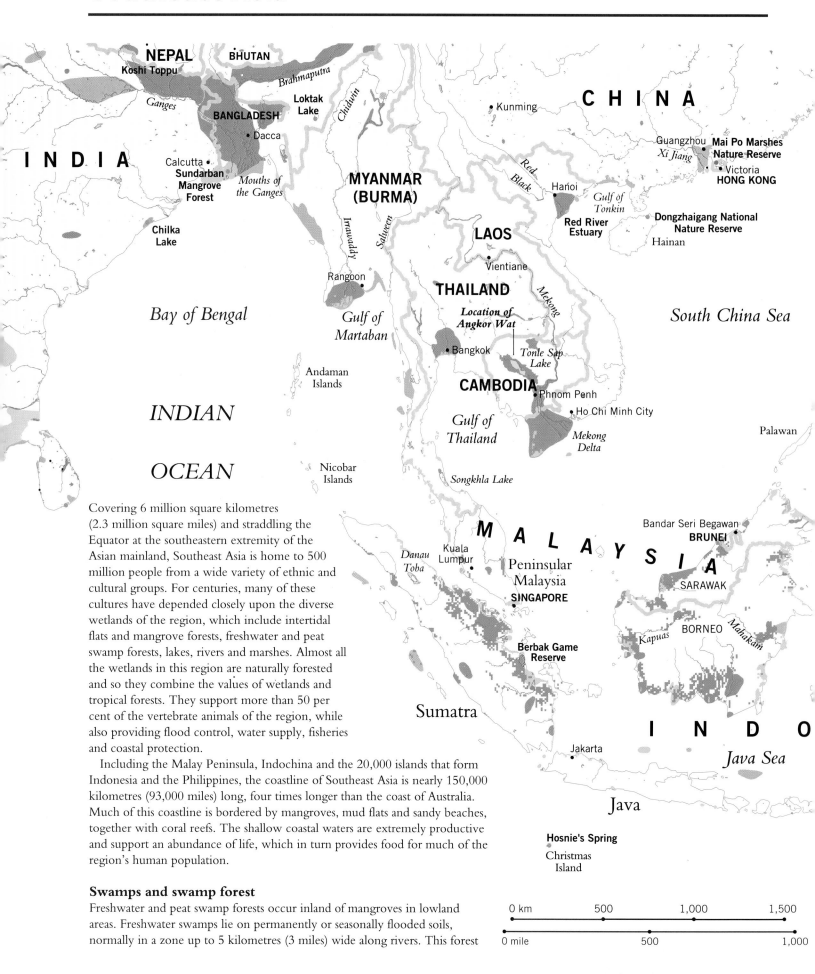

NEPAL
Koshi Toppu

BHUTAN

Brahmaputra

**Loktak
Lake**

Chidwin

CHINA

• Kunming

Ganges

BANGLADESH

• Dacca

I N D I A

Calcutta •
**Sundarban
Mangrove
Forest**

*Mouths of
the Ganges*

Chilka
Lake

**MYANMAR
(BURMA)**

Irrawaddy

Saluween

Guangzhou
Xi Jiang

**Mai Po Marshes
Nature Reserve**

• Victoria
HONG KONG

Red

Black

Hanoi •

*Gulf of
Tonkin*

**Red River
Estuary**

**Dongzhaigang National
Nature Reserve**

Hainan

LAOS

• Vientiane

Rangoon •

*Gulf of
Martaban*

Bay of Bengal

THAILAND

Mekong

**Location of
Angkor Wat**

*Tonle Sap
Lake*

South China Sea

• Bangkok

Andaman
Islands

INDIAN

CAMBODIA
• Phnom Penh

*Gulf of
Thailand*

• Ho Chi Minh City

*Mekong
Delta*

Palawan

OCEAN

Nicobar
Islands

Songkhla Lake

Covering 6 million square kilometres
(2.3 million square miles) and straddling the
Equator at the southeastern extremity of the
Asian mainland, Southeast Asia is home to 500
million people from a wide variety of ethnic and
cultural groups. For centuries, many of these
cultures have depended closely upon the diverse
wetlands of the region, which include intertidal
flats and mangrove forests, freshwater and peat
swamp forests, lakes, rivers and marshes. Almost all
the wetlands in this region are naturally forested
and so they combine the values of wetlands and
tropical forests. They support more than 50 per
cent of the vertebrate animals of the region, while
also providing flood control, water supply, fisheries
and coastal protection.

Including the Malay Peninsula, Indochina and the 20,000 islands that form
Indonesia and the Philippines, the coastline of Southeast Asia is nearly 150,000
kilometres (93,000 miles) long, four times longer than the coast of Australia.
Much of this coastline is bordered by mangroves, mud flats and sandy beaches,
together with coral reefs. The shallow coastal waters are extremely productive
and support an abundance of life, which in turn provides food for much of the
region's human population.

M A L A Y S I A

*Danau
Toba*

Kuala
Lumpur •

**Peninsular
Malaysia**

SINGAPORE

Bandar Seri Begawan •
BRUNEI

SARAWAK

BORNEO

Kapuas

Mahakam

**Berbak Game
Reserve**

Sumatra

I N D O

Java Sea

• Jakarta

Java

Hosnie's Spring

Christmas
Island

Swamps and swamp forest

Freshwater and peat swamp forests occur inland of mangroves in lowland
areas. Freshwater swamps lie on permanently or seasonally flooded soils,
normally in a zone up to 5 kilometres (3 miles) wide along rivers. This forest

0 km	500	1,000	1,500

0 mile	500	1,000

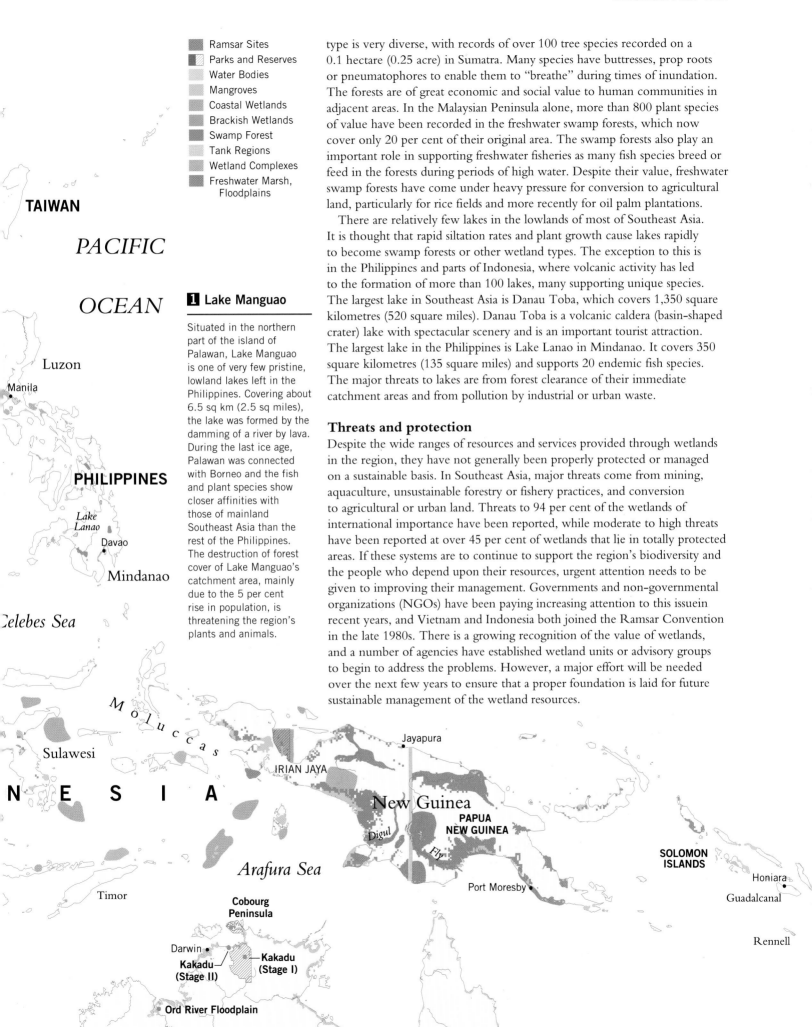

Legend

- Ramsar Sites
- Parks and Reserves
- Water Bodies
- Mangroves
- Coastal Wetlands
- Brackish Wetlands
- Swamp Forest
- Tank Regions
- Wetland Complexes
- Freshwater Marsh, Floodplains

TAIWAN

PACIFIC

OCEAN

Luzon

Manila

PHILIPPINES

Lake Lanao

Davao

Mindanao

Celebes Sea

M o l u c c a s

Sulawesi

I N D O N E S I A

Timor

Arafura Sea

IRIAN JAYA

Jayapura

New Guinea

PAPUA NEW GUINEA

Digul

Fly

Port Moresby

SOLOMON ISLANDS

Honiara

Guadalcanal

Rennell

Cobourg Peninsula

Darwin

Kakadu (Stage II)

Kakadu (Stage I)

Ord River Floodplain

Lakes Argyle and Kununurra

1 Lake Manguao

Situated in the northern part of the island of Palawan, Lake Manguao is one of very few pristine, lowland lakes left in the Philippines. Covering about 6.5 sq km (2.5 sq miles), the lake was formed by the damming of a river by lava. During the last ice age, Palawan was connected with Borneo and the fish and plant species show closer affinities with those of mainland Southeast Asia than the rest of the Philippines. The destruction of forest cover of Lake Manguao's catchment area, mainly due to the 5 per cent rise in population, is threatening the region's plants and animals.

type is very diverse, with records of over 100 tree species recorded on a 0.1 hectare (0.25 acre) in Sumatra. Many species have buttresses, prop roots or pneumatophores to enable them to "breathe" during times of inundation. The forests are of great economic and social value to human communities in adjacent areas. In the Malaysian Peninsula alone, more than 800 plant species of value have been recorded in the freshwater swamp forests, which now cover only 20 per cent of their original area. The swamp forests also play an important role in supporting freshwater fisheries as many fish species breed or feed in the forests during periods of high water. Despite their value, freshwater swamp forests have come under heavy pressure for conversion to agricultural land, particularly for rice fields and more recently for oil palm plantations.

There are relatively few lakes in the lowlands of most of Southeast Asia. It is thought that rapid siltation rates and plant growth cause lakes rapidly to become swamp forests or other wetland types. The exception to this is in the Philippines and parts of Indonesia, where volcanic activity has led to the formation of more than 100 lakes, many supporting unique species. The largest lake in Southeast Asia is Danau Toba, which covers 1,350 square kilometres (520 square miles). Danau Toba is a volcanic caldera (basin-shaped crater) lake with spectacular scenery and is an important tourist attraction. The largest lake in the Philippines is Lake Lanao in Mindanao. It covers 350 square kilometres (135 square miles) and supports 20 endemic fish species. The major threats to lakes are from forest clearance of their immediate catchment areas and from pollution by industrial or urban waste.

Threats and protection

Despite the wide ranges of resources and services provided through wetlands in the region, they have not generally been properly protected or managed on a sustainable basis. In Southeast Asia, major threats come from mining, aquaculture, unsustainable forestry or fishery practices, and conversion to agricultural or urban land. Threats to 94 per cent of the wetlands of international importance have been reported, while moderate to high threats have been reported at over 45 per cent of wetlands that lie in totally protected areas. If these systems are to continue to support the region's biodiversity and the people who depend upon their resources, urgent attention needs to be given to improving their management. Governments and non-governmental organizations (NGOs) have been paying increasing attention to this issue in recent years, and Vietnam and Indonesia both joined the Ramsar Convention in the late 1980s. There is a growing recognition of the value of wetlands, and a number of agencies have established wetland units or advisory groups to begin to address the problems. However, a major effort will be needed over the next few years to ensure that a proper foundation is laid for future sustainable management of the wetland resources.

The Mekong River

With a catchment area of 795,000 square kilometres (305,000 square miles) and stretching for 4,200 kilometres (2,600 miles), the Mekong is the longest river in Southeast Asia and the twelfth longest in the world. It originates in the Tibetan highlands on the southern border of Tsinghi Province, China. At the end of its long journey, the river drains into the South China Sea through its delta, which lies to the south of Ho Chi Minh City in Vietnam. The river is called Cuu Long ("Nine Dragons") by the Vietnamese after the delta's complex web of distributary channels. From just below Phnom Penh to the sea, the delta covers an area of 55,000 square kilometres (21,000 square miles), and is one of the largest and most important wetland systems in Asia.

Along its way to the sea, the Mekong is the lifeblood of the region. About three-quarters of its drainage basin lies within Laos, Thailand, Cambodia and Vietnam and about 40 per cent of the total population of these four countries live in the basin. The cities of Vientiane and Phnom Penh are located on its banks, and the river plays a central role in the economy of the whole region.

Hydroelectricity

Since 1957, at least 100 hydroelectric dams, including seven major dams, have been proposed for the Mekong and its tributaries. In addition to generating electricity, the proposed dams would help to exploit the estimated 39,000 square kilometres (15,000 square miles) of potential agricultural land in the delta by providing controlled irrigation water.

So far, just over a dozen dams have been finished and several others are under construction. While these have helped increase agricultural production, negative effects are beginning to be felt. Beneficial flooding during the monsoon season has been reduced. Freshwater inflow to the coastal ecosystems, including the mangrove forests, has decreased, and salinity patterns and levels have been altered. If more of the proposed dams are constructed it is expected that the adverse effects will increase. The species of fish that use the Mekong River system for spawning grounds and migration paths will be disturbed by dam construction upstream, a continued lack of freshwater inflow will severely degrade the mangrove forests, while a reduction in flood-borne silt will decrease soil fertility and agricultural production.

Tonle Sap Lake and Tonle Sap River

During the dry season, Cambodia's Tonle Sap (Great Lake) covers some 2,500–3,000 square kilometres (970–1,150 square miles). As the water level in the Mekong rises in June or July, the flow in the Tonle Sap River is reversed and the Mekong floodwaters enter Tonle Sap. At the height of the flood season in September and October, the lake and its inundation zone can cover as much as 13,000 square kilometres (5,000 square miles) of plains extending from the northwestern corner of the country to the Mekong at Phnom Penh. Large tracts of fresh water swamp forest grow in the floodplain and estimates in the 1970s put the total area of forest in Cambodia at about 6,800 square kilometres (2,600 square miles). But by 1990 almost 20 per cent of the swamp forest had been cleared for firewood, agricultural land and fish ponds.

Right **An area of paper-bark forest (*Melaleuca* sp). The tree gets its common name from the extremely thin layer of bark, which, as can be seen in the picture, easily peels away from the tree. The once extensive paper-bark forests of Southeast Asia were severely reduced in area during the course of the Vietnam War.**

Below **The Mekong River near Luang Prabang in northern Laos. In the relatively densely populated regions along the banks of the Mekong, every available area of land is used for agriculture. A variety of crops is grown on mini terraces to supplement the population's diet, with water being transported simply by watering can, or any other container, from the river to the bank.**

Below right **The Mekong River in Thailand is an important source of fish for hundreds of riverside communities. The simple bamboo structures standing in the water are where the fisherman hang their nets when not in use. The stilted houses show the local awareness of the danger of flooding.**

War and the Mekong Delta

Vietnam is facing one of its biggest challenges since the Vietnam War ended in 1975. American and Vietnamese scientists estimate that 22,000 square kilometres (8,500 square miles) of forest and a fifth of the country's farmland were affected as a direct result of bombing, mechanized land-clearing and defoliation. About 2,800 square kilometres (1,000 square miles) of the Mekong Delta was under mangrove and *Melaleuca* forest at the start of the war. From the early days of the war, the delta was extensively sprayed with defoliants, including Agent Orange, which resulted in the destruction of an estimated 1,240 square kilometres (480 square miles) of mangrove and 270 square kilometres (100 square miles) of *Melaleuca* forest. This amounted to more than 50 per cent of the entire mangrove forest in Vietnam.

The exposure of the alluvial forest soil to high temperatures after trees were defoliated caused desiccation and leaching of nutrients from the soil, making it very hard and unsuitable for plants. Some of the chemicals remained active in the soil for many months and subsequently destroyed the soil's micro-organisms. As a result, there has been little recolonization in many areas. Efforts by the Vietnamese have, however, improved the rate of recovery; more than 700 square kilometres (270 square miles) have now been replanted.

The *Melaleuca* forests that grew on the plains behind the mangroves were burnt with Napalm as well as sprayed. Furthermore, canals were dug to drain the flooded areas in another effort to flush out the guerrilla army. The dried areas became highly acidic due to chemical changes in the soil caused by oxidization and the land infertile; the *Melaleuca* forests were almost destroyed. Local and international efforts have been made to restore the forests and the fertility of the land by replanting. In the long-term, it is intended to harvest the *Melaleuca* for timber, but the short-term plan is to harvest honey from bees feeding on the rich flowers and to extract oils from the leaves.

Mangroves

Growing predominantly in Indonesia, Malaysia and Papua New Guinea, the mangroves of Southeast Asia cover more than 60,000 square kilometres (23,000 square miles) – about 1 per cent of the world's total area. Containing more than 60 species of tree, the mangrove forests in this region support a variety of animal life, including more than 40 species of mammal and 200 species of bird. The mangroves are also essential as the nursery and feeding areas of many of the region's commercial fish and prawn species.

Because much of the human population of Southeast Asia lives in coastal zones, mangroves have been under particularly heavy pressure from over-exploitation of their resources and from clearance for coastal aquaculture, agriculture and urban development. In the Philippines, for example, mangroves have been reduced in area by 75 per cent during the last 60 years. In Malaysia, proposals for large-scale coastal reclamation threaten the remaining forests on the west coast of the peninsula. In Thailand, the area of mangrove is estimated to have been reduced by 22 per cent, from 3,680 square kilometres (1,420 square miles) in 1961 to just under 2,900 square kilometres (1,120 square miles) in 1979. Clearance for aquaculture during the past decade has been especially rapid, and although figures are not yet available, it is feared that mangrove areas may have suffered a further devastating decline. Fortunately, most of the mangroves' associated mud flats still remain, although some have been lost to aquaculture and industry.

Protecting the mangroves

In Indonesia, significant steps are now being taken to conserve the dwindling mangrove resource. The legal status of Indonesia's mangroves is such that commercial harvest of the trees is regulated and requires harvesters to leave an undisturbed protection zone 100 times wider than the tidal range along the seaward margin and 50 metres (160 feet) wide along rivers. Subsistence use of mangroves is not regulated, although major projects are currently being developed to improve management of mangrove forests by involving local communities in replanting and developing systems of sustainable use. More than 30 protected mangrove areas have been declared by the Minister of Agriculture and several others have

been proposed. Similarly, in the Philippines a total of 780 square kilometres (300 square miles), 58 per cent of the remaining mangrove forest, have been identified for designation as conservation and preservation areas. And in Thailand, the Government is providing funds for rehabilitation of mangroves in areas where aquaculture has proved unsuccessful.

Above **The rare false gavial is one of the more endangered inhabitants of the Berbak Nature Reserve. Although the false gavial is classified with the Crocodylidae family, its long, narrow snout suggests that it is more closely related to the gavial (*Gavialis gangeticus*) found in parts of southern Asia.**

Right **The dense, green vegetation of the swampy, tropical forest of Berbak Nature Reserve on the east coast of Sumatra, Indonesia. The reserve supports a vast array of animal and plant species, and among the latter is found the richest assemblage of palms yet found in a swamp forest.**

Flooded Forest

Peat swamp forest covers about 200,000 square kilometres (77,000 square miles) of Southeast Asia, mostly in Indonesia and Malaysia. Figures for freshwater forest are less comprehensive, but Indonesia alone possesses 50,000 square kilometres (20,000 square miles). Peat swamp forest normally develops from freshwater swamp forests in which leaf litter and other organic debris has accumulated in layers of peat up to 20 metres (65 feet) thick. The peat swamp is acidic, often domed and supports a forest that is characteristically zoned in concentric bands around the top central dome. The vegetation varies from about 100 tree species in the outer mixed-forest zone to virtually single species forest, known as *padang*, on the top of the dome .

Freshwater swamp forests and peat swamp forests play an important role in the mitigation of flooding in adjacent areas by acting as natural reservoirs that absorb and store excess water during the rainy season. These types of swamp are also major forestry resources, with many valuable timber species. The principal threats to freshwater swamp forest are conversion to agricultural use and non-sustainable exploitation for timber. These problems are particularly acute in Malaysia, where freshwater swamp forest is probably the most severely threatened wetland habitat. Although peat swamp forest, unlike freshwater swamp forest, is of only marginal use for agriculture, clearance for aquaculture ponds is a major threat, together with non-sustainable logging. Under current practice, it is predicted that all of Sarawak's peat swamp forest will have been logged before the year 2000.

Berbak Nature Reserve

Berbak Nature Reserve in Jambi Province on the east coast of Sumatra covers an area of approximately 1,650 square kilometres (630 square miles). It consists mostly of peat swamp forest and contains an amazingly diverse flora. The forest also supports a remarkably diverse mammal fauna, including Sumatran tigers (*Panther tigris sumatrae*), clouded leopards (*Neofelis nebulosa*), Malayan tapirs (*Tapirus indicus*) and Sumatran rhinoceroses (*Dicerorhinus sumatrensis*). More than 160 species of bird are known to live in the reserve, and it may be one of the last remaining strongholds of the rare crocodilian, the false gavial (*Tomistoma schlegelii*).

The swamp forests of Berbak are thought to supply a wide range of benefits, including the support of fisheries, protection against saline inundation of inland agricultural areas and flood protection. The Asian Wetland Bureau and the Directorate General of Forest Protection and Nature Conservation of the Indonesian Government have recently completed a detailed study of the management needs of this area. The Berbak Nature Reserve is currently under severe threat from illegal logging and clearance, poaching and development. It was, however, declared as the country's first Ramsar Site when Indonesia ratified the Convention in 1992, and this should encourage greater support for conservation efforts in the area.

Above **Mangroves are often used on building sites in many parts of Southeast Asia. Mangrove forests are being reduced on a vast scale in this region to make way for agriculture, aquaculture and urban developments. The cut mangroves are a much** cheaper building material than expensive steel scaffolding. The large-scale destruction of mangroves in Southeast Asia has prompted some governments to introduce strict laws regulating the amount of mangrove forest that can be cleared.

Australia

Covering 7.7 million square kilometres (3 million square miles), the Australian continent is vast, mostly flat and much of it extremely arid. Yet, when it rains, floods spread across large areas, contrasting vividly with the prevalent dryness. The flatness and the extremes of flood and drought are major factors in governing the distribution and type of wetland over much of the continent. Permanent lagoons, swamps and marshes lie in areas of higher rainfall, while in the arid zones wetlands may seem to be non-existent for years, only to temporarily dominate the landscape after heavy rains. The flora and fauna of these wetlands are those that respond to episodes of plenty, and their ecology is overwhelmingly one of abundant life and widespread death.

Australia is, in geological terms, very old. Erosion and sedimentation have created extensive landscapes of extremely low relief; Lake Eyre, into which drains much of the central interior, is many metres below sea-level. Because of its geology, huge areas of inland Australia generate salts that accumulate in saline lakes which may be dry for years.

Australia's varied wetlands

Although essentially arid, some areas of Australia support a surprisingly rich variety of wetlands. Short, perennial streams drain across narrow coastal plains; large, biologically rich estuaries lie around the northern and eastern coastlines, while west of the Great Dividing Range, which almost joins Cape York Peninsula in the north to the southern margins of the continent, huge river systems drain into the interior. The Murray-Darling river system drains about 14 per cent of the continent.

Wetlands lie along the courses of the drainage systems in the form of swamps, floodplains and stream beds. These are often dry for many months or years, but during floods they change dramatically, and parts of inland Australia may suddenly be awash with swirling, turbid water. Floodwaters spill out from the drainage channels, changing the parched landscape into a kaleidoscope of briefly flowering plants. After the floods, shallow inland lakes, which gradually recede over months and sometimes years, remain as oases teeming with fish, birds and marsupials feeding on the rich invertebrate and plant life. These wetland ecosystems are so significant that they influence the migrations of many birds, including shorebirds and seabirds that move inland in these times of plenty.

Wetland species

Where wetlands are more sustained or predictable, they support forests. In freshwater areas these forested wetlands are dominated by species of eucalyptus, such as the coolibah (*Eucalyptus microtheca*) and the river red gum (*Eucalyptus camldulensis*), paper-bark trees (*Melaleuca* spp.), she-oaks (*Casuarina* spp.) and the freshwater mangrove (*Barringtonia* spp.).On the floodplains of the Murray River, *Eucalyptus camaldulensis* grows in sufficiently large quantities to constitute a significant hardwood forest resource.

Both Northern Territory and Queensland have extensive freshwater floodplains covered by grasses, such as wild rice (*Oryza meridionalis*) and spiny mudgrass (*Pseudoraphis spinescens*), and tall trees, such as paper-barks. Reeds, sedges, water lilies and herbs abound in a diverse floral display.

Cobourg Peninsula

Timor Sea

INDIAN

OCEAN

Kakadu (Stage II)
Darwin
Rum Jungle

Ord River Floodplain

Lakes Argyle & Kununurra

Roebuck Bay

Eighty-mile Beach Reserve

Lake Mackay

WESTERN AUSTRALIA

Shark Bay

Perth
Forrestdale & Thomsons Lakes

Peel-Yalgorup System

Lake Toolibin

Vasse-Wonnerup System

Lake Warden System

Great Australian

Ramsar Sites
Protected Areas
Mangroves
Salt Lakes, Salt Pans
Seasonally Flooded Wetlands
Occasional Wetlands

Arafura Sea

Murgenella
Cooper

Kakadu National Park
Arnhem
Land

uth Alligator

Cape York

Great

Gulf of
Carpentaria

Barrier

NORTHERN
TERRITORY

PACIFIC

OCEAN

Cairns

Reef

Townsville

Great

• Alice Springs

QUEENSLAND

Diamantina

Dividing

Dartmouth Lake

Range

Macumba

Coongie Lakes

Moreton
Bay

Brisbane •

Gold
Coast

Neales

Cooper Creek

Lake Eyre

SOUTH
AUSTRALIA

Darling

Macquarie Marshes
Nature Reserve

NEW SOUTH
WALES

Koonagang
Nature Reserve

Bight

Riverland

Hattah-Kulkyne Lakes

Sydney •

Towra Point
Nature Reserve

Adelaide •

Murray

Lake
Albacutya

Barmah
Forest

Canberra •

Kerang Wetlands

Hume Reservoir

The Coorong and Lakes
Alexandrina and Albert

Gunbower Forest

VICTORIA

Bool and Hacks Lagoons

Melbourne •

Tasman
Sea

Western District Lakes

Western
Port

Gippsland Lakes

Corner Inlet

Port Phillip Bay
(western shoreline)
& Bellarine Peninsula

Sea Elephant
Conservation Area

Logan Lagoon

Little Waterhouse Lake

Cape Barren Is.
East Coast Lagoons

TASMANIA

Jocks Lagoon
Lower Ringarooma River
Apsley Marshes
Moulting Lagoon

Lake Crescent
(northwestern corner)

Pittwater-Orielton
Lagoon

Hobart •

The animal life of these seasonally flooded ecosystems is no less diverse, with the vertebrates being far better known than the myriad invertebrate species. Freshwater and saltwater crocodiles (*Crocodylus porosus* and *C. johnstoni*), turtles, such as the pignosed (*Carettochelys insculpta*), and fishes, such as the barramundi (*Lates calcarifer*), are numerous.

Saline wetlands

Mangroves and salt marshes are common along much of the long northern and western coastline, with the total area of mangrove in Australia estimated at over 11,600 square kilometres (4,500 square miles). Outside Indonesia, this is the largest area remaining in the countries of Southeast Asia and the Pacific. Pressure for coastal development has led to increasing loss of mangrove forest in many parts of Australia.

Growing awareness of their value is, however, leading to action. In Moreton Bay in southern Queensland, for example, it has been estimated that commercial and recreational fisheries based on mangrove-dependent species are worth about 280 million Australian dollars annually. This has resulted in the founding of a marine park, which in turn has halted the expansion of Gold Coast City, the main cause for mangrove destruction in Moreton Bay.

Map Note

Australia is a very dry continent but its interior is characterized by regions which flood periodically and become important wetlands. The reasons for the flooding and the periodicity determine the character of the wetland. Those which flood annually are shown as seasonally flooded and those which flood when they are subject to irregular rainfall are shown as saltpans/seasonal wetlands. A third category, shown as "occasionally flooded", are known locally as the Channel Country and depend on rainfall to the north and east which floods irregularly into an otherwise arid region.

Aboriginal Use of Wetlands

Long before Captain Cook set foot on Australian soil in 1770, the wetlands of northern Australia played a central role in the lives of many of Australia's aboriginal peoples, and they continue to do so today. Coinciding with the formation of the freshwater wetlands of northern Australia, between 500 and 1,200 years ago, there was a sudden increase in human occupation of the plains of the Alligator Rivers Region in the Northern Territory. People here are thought to have followed a pattern of seasonal nomadism, congregating on the wetland margins in the dry season and moving to higher ground in the wet season.

Today, the Australian Aborigines still use the food resources of wetlands, although the ready supply of food from feral buffalo has reduced the use of natural wetland foods. These buffaloes also degrade the wetlands and this has combined with saltwater intrusion, due to overstretched groundwater reserves used for irrigation, and introduced-plant invasion to reduce the amount and number of types of food available.

Aboriginal women harvest plant foods from freshwater wetlands during the dry season, when the food is more easily accessible. Hunting and fishing are mainly done by the men.

Lake Eyre Basin

Lake Eyre Basin is one of the largest internal drainage systems in the world. It covers 1.1 million square kilometres (425,000 square miles) and is underlain by the world's largest artesian basin. At its centre, the Lake Eyre Basin is generally arid, with average annual rainfall between 100 and 150 millimetres (4 and 6 inches). The wetland systems the basin contains form oases in a vast stone and sand desert. The oases are among the most dynamic and spectacular in Australia. They are fed by the four major rivers that drain into Lake Eyre, namely the Cooper and Diamantina in the east, and the Neales and Macumba in the west.

Rainfall is erratic in timing, location and intensity, so the discharge of even the large rivers is extremely unpredictable. Yet it is this variability, combined with the low gradients and the relief of the basin, that is responsible for the complexity of the Lake Eyre wetlands. Plants and animals respond with opportunism to the instability of the wetland environment. Plants flower and fish reproduce after floods, and waterbirds congregate to feed and breed,

generating some of the highest waterfowl densities in the world. In droughts, wetland animals emigrate, aestivate (lie dormant) or, like fish in the contracting lakes, die in multitudes. In such an environment, the permanent wetlands are vital refuges for the less mobile species. The mound springs in the south of the basin are fed by artesian waters, and provide islands of permanence; an inland Galápagos, they are rich in unique species.

Growing threats

Because the rivers have not been substantially modified or polluted, the wetlands are still relatively pristine, and no significant wetland loss has occurred. However, this situation is now changing. The whole basin is subject to intensive grazing pressure from rabbits and cattle and much of it is being dissected by mining exploration tracks. In addition, the acacia woodlands of the headwaters are being cleared, and damage from four-wheel-drive tourism is increasing rapidly.

Kakadu National Park

A report to Unesco's (United Nations Educational, Scientific and Cultural Organization) World Heritage Committee stated that: "There is simply no other protected area on Earth like Kakadu." Stretching over 20,000 square kilometres (7,700 square miles) and including almost the entire catchment area of the South Alligator River, the park is the largest in Australia and includes a tremendous diversity of habitats and species found nowhere else in the world. Located in the north of the Northern Territory, Kakadu's wetlands are listed under the Ramsar Convention, and include mangroves on offshore islands, along the coast and fringing the rivers, salt flats, and creeks that overflow onto extensive freshwater wetlands, grasslands, sedgelands and swamps. The biological value of these wetlands is enormous. The vegetation is species-rich with, for example, 22 mangrove species and 225 freshwater plants on the Magela floodplain. The birdlife is prolific, with peak populations reaching 3 million individuals, 85 per cent of them magpie geese (*Anseranas semipalmata*). Up to 400,000 wandering and 70,000 plumed whistling ducks (*Dendrocygna eytoni*), 20,000 Radjah shelducks (*Tadorna radjah*), 50,000 Pacific black ducks (*Anas superciliosa*) and 50,000 grey teal also use the area. Thirty-five species of shorebird, many of them migrants from the Arctic, use these wetlands as well. The wetlands near the coast support turtles and the estuarine crocodile. Other turtles, snakes and the freshwater crocodile also live in the freshwater wetlands, along with seemingly innumerable frogs.

Change and development

Management problems centre on introduced animals and plants, tourism and mining operations. Buffalo were a major problem on the freshwater floodplains, but have been virtually eliminated. Feral pigs, however, are a serious problem and extraordinarily difficult to control. As for vegetation, the most serious threat is posed by the prickly mimosa shrub. To date, it has been kept at bay in the park, but severe infestations on the park's boundaries make this species a potential problem. Increasing numbers of tourists have been visiting Kakadu over the last decade. Balancing the needs and expectations of these visitors with nature conservation priorities is an ongoing process. The issue that has generated most concern and controversy is the development of uranium mining in the Magela Creek catchment area.

Left Paper-bark trees dominate this area of freshwater swamp in the Kakadu National Park in Australia'a Northern Territory. Other wetland habitats that are represented within this famous protected area include: mangroves, salt flats, billabongs (small stagnant pools), floodplain grassland, sedgelands, and swamp herbfields. It is this wide variety of habitats that is chiefly responsible for the species diversity within Kakadu National Park.

Below A pair of white-breasted sea-eagles (*Haliaeetus leucogaster*) on an over-water perch in Kakadu. These eagles are ferocious predators, and will even take the large brogla (*Grus rubicunda*).

Bottom This view of Dillinna Creek south of Lake Eyre emphasizes the flat landscape and the temporary nature of the region's wetlands. Although the creek is full of water, sand has collected around the plants – evidence of drier conditions.

New Zealand and the Pacific

The volcanic origins and glaciated history of New Zealand combine with today's maritime climate and heavy rainfall to produce a diverse wetland landscape. Features include such things as rivers and bogs from frequent rain, glaciation and volcanic action, swamps from the deposition of erosion products by rivers and the sea, and estuaries and lagoons from tidal flooding of old Pleistocene (Ice Age) valleys.

The wetlands are widespread and diverse. They support distinctive communities, which contribute to the unique biological and geographic character of New Zealand. They include swamps of *Phormium* spp., braided rivers, which form good waterfowl habitat, bogs, saline rush and reed estuaries with *Leptocarpus* spp. and *Juncus* spp., and kahikatea (*Podocarpus dacrydioides*) swamp forest.

The distribution of these wetlands reflects a combination of geological history, relief, climate and the intensity of wetland modification. For example, freshwater wetlands occupy only a few square kilometres in the eastern North Island, while in South Island they occupy some 300 square kilometres (120 square miles). While of the 300 or so estuarine wetlands distributed around the country's coastline, some of the largest systems occur in the northern end of North Island.

Conversion and loss

There is a long history of use and modification of wetlands in New Zealand. Land drainage, gold mining, flood control, land clearance, general agricultural development, kauri-gum digging and flax milling have all contributed to wetland loss. It is estimated that only 10 per cent of pre-European wetlands remain, and wetlands now occupy less than 2 per cent of the total land area of New Zealand.

The process of wetland loss has fragmented the wetland resource, and conservation action needs to be increased in the face of continuing pressure from extraction of sand and gravel, reclamation of estuaries, lagoons, lake shores and river margins, flood control, pollution and land drainage. Priorities include the establishment of buffers of indigenous vegetation along rivers and around the margins of lakes, swamps and estuaries, and the protection of corridors linking wetlands of all kinds to other protected ecosystems, both terrestrial and marine.

The Pacific islands

Perceived by many as paradise, the total land area of the 23 small island nations and territories of the Pacific region is merely 107,000 square kilometres (41,000 square miles), rather less than that of North Island, New Zealand. Thus, while these islands support a wide variety of wetland types, most are very limited in their extent because of the tiny size of most of the islands. Yet, despite their small size, these wetlands yield a variety of resources, some of which are used intensively by the island peoples. Many islands, for example, provide the basis for sustainable *taro* agriculture, a crop grown for its edible root.

By far the most extensive wetlands are the shallow lagoons and reef flats that fringe many of the larger islands and comprise the bulk of most of the low-lying atolls (ring-shaped coral reefs surrounding a central lagoon and with breaches to the open sea). Mangroves also occur widely in the western and northwestern Pacific region.

Freshwater wetlands, such as lakes, marshes, swamps and bogs, are well represented on the larger islands of New Caledonia, Fiji and the Solomon Islands, but are very limited on the smaller islands, and virtually non-existent on atolls. Most of the freshwater lakes lie on mountainous volcanic islands, where they have formed in extinct or dormant volcanic craters. Examples include Rano Kau, Rano Raraku and Rano Aroi on Easter Island, Lake

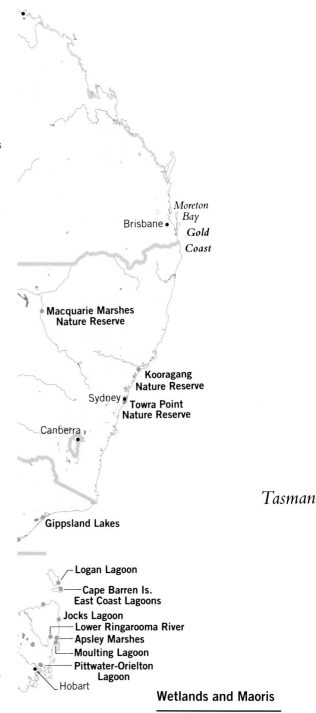

Brisbane

Moreton Bay

Gold Coast

Macquarie Marshes Nature Reserve

Kooragang Nature Reserve

Sydney

Towra Point Nature Reserve

Canberra

Tasman

Gippsland Lakes

Logan Lagoon

Cape Barren Is. East Coast Lagoons

Jocks Lagoon

Lower Ringarooma River

Apsley Marshes

Moulting Lagoon

Pittwater-Orielton Lagoon

Hobart

Wetlands and Maoris

Wetlands have always had an important role to play within Maori communities – in an intricate inter-connected relationship that links wetlands with the people in both a cultural and material sense. Wetlands have provided food; plants for weaving, medicines, dyes; canoe landing sites; a place to season timber; and been used to store artifacts.

Banks
Islands

VANUATU

Port Villa

New
Caledonia *Plaine des Lacs*

Savaii Upolu
WESTERN
SAMOA

FIJI

Suva

PACIFIC

TONGA

OCEAN

■ Ramsar Sites
Water Bodies
Mangrove
Estuaries, Coastal Wetlands
Pans, Seasonal Wetlands
Swamp Forest
Freshwater Marsh, Floodplains

0 km   500   1,000

0 mile   250   500

Waiau on the Big Island in Hawaii, Lagona Lake and Inner Lake on Pagan in the northern Marianas, the Lakes of Niuafo'ou and Tofua in Tonga, Lake Letas in the Banks Islands (the largest freshwater lake in the Pacific islands) and Lanioto'o Lake on Upolu in Western Samoa. Many of these lakes have special cultural importance for the local people.

Extensive areas of swamp or marsh are rare on the Pacific islands, notable exceptions being Kawainui Marsh on Oahu (the largest freshwater swamp in Hawaii) and the Plaine des Lacs, a complex of freshwater lakes and marshes covering 200 square kilometres (75 square miles) near the east end of the island of New Caledonia. Although extensive areas are rare, there are small areas of freshwater marsh and swamp on most of the larger islands. Typically, these wetlands are fringed with tall stands of the reed *Phragmites karka*. Almost wherever they occur, these wetlands have been modified for the production of *taro*. Extensive *Sphagnum* bogs are confined to high elevations, usually over 1,500 metres (5,000 feet) on the larger Hawaiian Islands, such as Alakai Swamp on Maui.

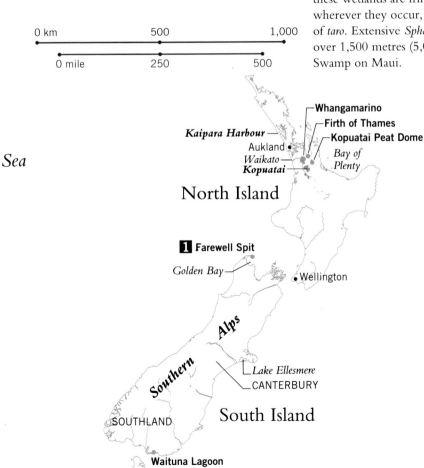

Sea

Whangamarino
Firth of Thames
Kopuatai Peat Dome
Kaipara Harbour
Aukland
Bay of
Waikato
Plenty
Kopuatai

North Island

■ Farewell Spit

Golden Bay
•Wellington

Alps

Southern

Lake Ellesmere
CANTERBURY

SOUTHLAND   South Island

Waituna Lagoon

■ Farewell Spit

Farewell Spit is a unique, 30-km (18-mile) long sandspit lying at the north of Golden Bay, South Island. It covers a land area of some 20 sq km (8 sq miles). The spit is estimated to have originated 6,500 years ago, derived from material eroded from the Southern Alps and West Coast seacliffs. Because the southern end of the Spit has extensive tidal sand and mud flats it provides feeding areas or habitats for 83 species of wetland bird, both national and international waders, including knots, eastern bar-tailed godwit, turnstone and South Island pied oystercatcher (*Haematopus ostralegus finschi*) and is a major overwintering area for an estimated 12,000 black swans. In 1976, the Spit was listed under the Ramsar Convention.

New Zealand's Birdlife

Of all habitats within New Zealand, wetlands support the greatest concentration of bird species. It is estimated that less than 2 per cent of the land area is wetland, yet of the birds which are regular visitors or are permanently resident and breeding, 22 per cent (over 50 species) have wetlands as their primary habitat, and a further 5 per cent have wetlands as important secondary habitat. 158 species (133 native) have been recorded at Lake Ellesmere; 80 of them are regular users of the lake. This number can be compared with, say, 20 species in a comparable area of native forest.

Human Impact

In a region where flat land suitable for agriculture and development is very much at a premium, wetlands in the Pacific are often seen as wastelands, easily converted to other uses. This situation is further exacerbated by the rapid growth in population. Because of their small size, the freshwater wetlands of the Pacific are extremely vulnerable to modification or destruction, and many natural marshes have now been drained for agriculture and other forms of development.

Coastal wetlands and mangroves have frequently been modified or destroyed by infilling to provide land for housing, industry, airports and harbours. Similarly, because of the restricted development of mangrove swamps in the Pacific and their often stunted growth forms, the value of mangroves has not been widely appreciated and little legislation exists for their protection. In Fiji, large areas of mangrove swamp have been destroyed for shrimp production and the cultivation of sugar cane and rice, while in Western Samoa, mangroves are being reclaimed for settlement or used as rubbish dumps.

Polluted waters

Massive amounts of sediment, transported downstream as a result of poorly managed forestry, slash-and-burn agriculture and mining activities, have led to the siltation and destruction of many coastal wetlands. Seagrass beds and coral reefs near the shore are particularly affected by this problem. The discharge of urban and industrial effluents has caused serious pollution in wetlands and coastal waters, and wetlands have all too frequently been used for the disposal of rubbish and waste oils. On many of the more densely populated islands, especially in Micronesia, high levels of pollution have resulted in the mass extinction of stream floras and faunas. The use in prawn fishing of chlorox to poison streams has been particularly harmful on some islands. In French Polynesia, the main rivers on Tahiti (Papenoo and Punaruu) are now badly degraded by mining for sand and gravel.

Despite the diversity of these pressures, their impact is often only localized, and there are fine examples of natural wetlands on many of the larger islands. Most of the governments in the region are becoming more aware of the need to use these valuable resources wisely, and in several countries, coastal zone management plans have been developed which take into consideration the need for wetland conservation.

Lakes of the Pacific Region

Because of their extreme isolation, many of the freshwater lakes, marshes and streams in the Pacific islands have various rare and endemic species of flora and fauna. Streams in American Samoa and on several of the Micronesian islands, for example, are home to fishes that are found nowhere else in the world. There is at least one endemic species of fish, *Neogallaxia neocaldonicus*, confined to Lac en Huit in the Plaine des Lacs in New Caledonia, and an endemic sea krait, *Laticauda cruckeri*, is confined to Lake Te-Nggano on Rennell Island in the Solomons.

About 17 species of waterbird are endemic to the Pacific islands, but seven of these are now believed to be extinct (the Mariana mallard, four species of rail and crake, the Samoan woodhen and the Tahitian sandpiper), and all the others are under threat. In some cases the threat to species is a direct result of wetland destruction, but in others it is primarily caused by introduced predators. One species, the Laysan teal, is confined to a single brackish pond on Laysan Island in the northwest chain of the Hawaiian islands.

Endemic plants include a species of tree in the swamp forests of Yap and Palau, and a palm, the ivory nut palm, in the swamp forests of Pohnpei and Chuuk in Micronesia.

Above **Forest vegetation and freshwater stream at Karamea in the northwest of South Island, New Zealand. The distribution of New Zealand's freshwater and saltwater wetlands is markedly different between the two islands. Whereas North Island contains the great majority of the country's estuarine wetlands, almost all freshwater wetlands are found on South Island.**

Right **Christmas Island, Kiribati. A low-lying atoll, the island contains many shallow brackish lakes, but the most important wetlands (and the most extensive) are those that are purely marine.**

Kopuatai Peat Dome

The Kopuatai Peat Dome, a Ramsar Site on the Ilauraki Plain, covers 9,665 square kilometres (3,730 square miles) of the largest raised bog left intact in New Zealand. It is also the largest lowland bog dominated by the giant "restaid" rush, the greater wire rush *Sporodanthus traversii,* and the only significant unaltered restaid bog in the world. The peat began developing 13,500 years ago, and at present the depth of the peat is up to 12 metres (40 feet) towards the centre. The hydrological regime of the peat is dominated by rainfall, receiving little nutrient-rich groundwater. As a result, the peat is acidic and low in nutrients. In contrast, in the surrounding swamplands, where occasional flooding leads to mineralization of the soil, nutrient levels are much higher and the system more productive.

Nine species of threatened plants and animals depend on the dome, including the endemic black mudfish (*Neochanna diversus*), the Australasian bittern (*Botaurus poiciloprilus*), North Island fernbird (*Bowdleria punctata vealeae*), banded rail (*Rallus philippensis assimilis*), marsh crake (*Porzana pusilla affinis*) and spotless crake (*Porzana tabuensis plumbea*). Threatened plants that find a refuge on the dome include the endemic greater jointed rush (*Sporadanthus traversii*), which covers 22 square kilometres (8 square miles), three bladderworts, the creeping club moss and a fern.

Because of the fragile nature of the peatland ecosystem, entry without a permit is prohibited. However, gamebird hunting, grazing and a small area of agriculture are allowed in the mineralized fringe. Expansion of agriculture is at present the most severe threat to this fringe.

Autumn leaves bring a blaze of glory to this lakeside woodland in Saskatchewan. Canada has innumerable lakes, and in certain regions there are as many as 30 for every 100 sq km (40 sq miles). Most of the lakes were originally formed with meltwater during the retreat of the glaciers. Many of the small lakes are rimmed with marsh, and are slowly filling with peat to become bogs. The resulting landscape is a mosaic of woodland, wetland and water.

The Challenge of Conservation

Over the past 500 years, since the Columbian "discovery", the industrialized world has prospered on the riches of the tropics. Today, international trade continues to exert significant pressure upon natural ecosystems worldwide. The lands and peoples encountered have paid the price for the prosperity that this global economic order has brought to many – landscapes have changed beyond recognition, and animals and plant species, as well as cultures, have been driven to extinction.

Growth and Development

Today, the world is at a very different turning point. After five centuries of ever-increasing exploitation of our planet's resources and of degradation of the environmental systems that sustain human life, governments met in Rio de Janeiro in June 1992 to debate our future. The occasion was the United Nations Conference on Environment and Development.

Columbus' voyage was, to a large extent, a leap into the unknown: in contrast the Heads of State who met in Rio de Janeiro had before them a distressingly clear vision of the future. They met in response to mounting evidence that unless action is taken almost immediately to address the growing list of environmental problems, human society faces declining agricultural and industrial productivity, rising poverty and social unrest, and the loss of many varied habitats and ecosystems.

All ecosystems have an important role to play in helping society meet the needs of the 21st century, but the productivity of wetlands and the multiple benefits that they yield give them a special significance. Similarly, the heavy burden people place on wetlands and their vulnerability to climate change highlight the need for effective management practices.

Below **Columbus landing in America, as depicted by a 17th-century artist. European colonists brought with them a self-confident approach. They saw the natural landscape as** something to be tamed. Whereas the native peoples had worked with the landscape, the European approach was to change the land through "improved" drainage.

Population growth

The world today is characterized by great social, political, economic and environmental change. Some of these changes have brought benefits, but for many people conditions are much worse than 25 years ago. Rising population, largely in the poorer regions of the world, is one of the major factors that will continue to weigh heavily upon environmental debate.

The world's population was estimated at 5.2 billion in 1990; a population of around 8.5 billion is predicted for 2025. This increase will be concentrated in the developing world; 60 per cent will be living in coastal zones and around half will live in towns and cities. As populations grow, pressure upon all natural resources undoubtedly will increase. Because of modern demographic changes, this pressure will be especially acute in coastal regions and near urban centres. Coastal wetlands and the floodplains of rivers are therefore likely to be increasingly damaged by urban, industrial and agricultural uses. Wetland conservation plans need to address these trends and their potentially damaging implications.

In many regions, the continued growth of human population is driven by intense poverty. Effective and socially equitable development is therefore essential to a successful long-term strategy for stabilizing the world's

Above **Floating marketplace in Bangkok. Six thousand years ago it was the productivity of floodplain wetlands in Africa and Asia that permitted a rapid population increase, which in turn stimulated civilization and urban development. Today, the process threatens to work partly in reverse. The rapid growth of civilized urban populations, which require living space, food and electricity, threatens the continued existence of many natural wetlands.**

Below **The construction of a new dam across the Manicouagan River in Quebec, Canada. The introduction to the region of hydroelectricity during the first half of the 1900s led to the rapid industrial development of Quebec, as it did in many other parts of the world. Increased population and production now requires additional power plants. Although hydroelectricity does not carry the pollution hazards of fuel-burning power stations, there may be hidden costs in terms of wetland destruction.**

population. The importance of wetlands in the lives of many rural communities gives them a central role in efforts to establish effective approaches to rural development.

Uninformed development

Development investment is still dominated by economists and engineers, for many of whom environmental concerns remain peripheral rather than urgent issues. If they are to change their policies, they will need to be convinced of the real value of wetlands and of the way in which wetland management can underpin rural development. As we seek to establish suitable models of management, much is being learned from traditional forms of wetland use. However, these practices must be adapted to today's conditions and supported by government policies that give greater control over resources to local communities. This in turn requires governments to understand the full value of wetland ecosystems and the role that their sustainable use can play in achieving social and economic goals.

Economic growth

It is widely predicted that in the next 30 years economic growth will make a major contribution to the quality of life and to the stabilization of population in areas of East and Southeast Asia, South and Central America and other regions. Economic growth, however, will continue to place enormous pressures on the natural resources on which sustainable development depends.

All too often, economic growth is based on the conversion of natural ecosystems for the short-term profit of a few entrepreneurs. Not only does this lead to a loss of diversity, but also to lost opportunities for local people who, for example, find their traditional fisheries and farm lands given over to intensive, yet unsustainable, shrimp production and cereal farming. To address this problem there must be effective policies that will not only give vigorous protection to those areas of highest ecological value, but will also help to identify those areas where economic expansion can be pursued without harming natural resources.

With rise in population and pursuit of economic growth, problems such as pollution have also increased. In the industrialized world, many of these problems have already reached crisis level, precipitating concerted programmes of action by governments at national and local levels, as well as many pressure groups. These programmes are costly, however, and the challenge will be for developing nations to address the problems before they become critical.

Climate Change

World temperature is predicted to rise above the 1985 average by about 1–2°C (2–3.5°F) by 2025, with temperature increases in the polar zones between two-and-a-half and five times the global average. Rainfall patterns are likely to change, with precipitation predicted to be considerably greater in northern high latitudes, and higher than average throughout much of the world. In the tropics, climatic extremes are predicted to increase in their severity. Monsoons and tropical storm systems may become more intense, and it is thought that in arid areas increased evaporation caused by rising temperatures will coincide with reduced rainfall.

It is clear therefore that while there continues to be substantial uncertainty over the precise pattern of climate change and rainfall distribution, the consensus is that wet areas will get wetter and dry areas dryer. Floods and droughts are likely to increase in number as well as in severity and duration. The world has spent billions of dollars in building structures to protect people and property from floods and to provide water in times of drought. At the same time, it is counting the enormous cost of this investment, and seeing the limited success of flood- and drought-control programmes in many countries. The prospect of further investment in flood and drought management is daunting. Action must be taken, however, and natural wetland ecosystems have much to contribute.

Using water

Climate change will affect hydrological cycles, and this increases the urgency of determining the limits of water extraction from river systems for irrigation and other purposes. Extraction schemes have had devastating effects upon many floodplains and estuarine wetland systems, often depriving rural people of their livelihoods. Studies of the environmental consequences of structural-engineering approaches to water-management help build awareness and need to be continued, but they must now be accompanied by more modelling of natural hydrological systems. These studies should be complemented by social and economic surveys of communities using the ecosystems downstream from projects. The studies can then be used to help design dams and other engineering structures that have a minimal impact upon river flow, and do not disrupt agricultural, fishery and pastoral activities, and minimize loss of biological diversity.

Rising sea levels

If current predictions are correct, the incidence and severity of tropical storms will also increase with the gradual rise in the temperature of the Earth's atmosphere. In response, some countries will build strengthened coastal defences, but the high cost of such projects will preclude this approach for many others. Instead, it will be necessary to adopt approaches that aim to concentrate new infrastructure in places less subject to damage, and to provide shelters for use by coastal communities. The role of coastal wetland systems, especially areas of mangroves, in helping to dissipate the force of storms and reducing erosion needs to be assessed more comprehensively and incorporated in coastal protection strategies and plans.

The threat caused by coastal storms will be increased significantly by the widely accepted projected rise of 17–26 centimetres (6.5–10 inches) in sea level by the year 2030. The need for immediate action, therefore, is especially acute.

Above **An emperor goose (*Anser canasicus*) drinks from an ice-rimmed pool in Alaska. Like many other waterbirds, emperor geese nest on the Arctic tundra. One of the likely consequences of global warming is an extension of the boreal forest northward into what is now tundra. This encroachment by the forests will engulf many waterbird breeding sites, and populations could decline dramatically.**

Right **A dead acacia about to be engulfed by Saharan dunes near Adrar Chiriet, Niger. Trends in global climate change suggest desertification is likely to increase so threatening many desert oases.**

The threat caused by coastal storms will be increased significantly by the widely accepted projected rise of 17–26 centimetres (6.5–10 inches) in sea level by 2030. The need for the above actions, therefore, is especially acute. In addition, sea-level rise will have a number of other consequences for coastal wetlands. Of special concern is the impact upon mangrove ecosystems, many of which will be displaced inland as the sea rises. The capacity of mangroves to adapt to these changes will be determined by the pace at which change in sea level occurs, by the presence of suitable terrain inland for the mangroves to colonize, and by the continued supply of an adequate volume of fresh water.

Changing communities

The wetland communities with which we are familiar today will change with the warming of the world's climate. While the precise pattern of these changes will be modified by the changes in rainfall and by a rise in sea level, there are a number of general patterns that seem likely to emerge. For example, mangroves will grow at higher latitudes, and the range of temperate salt marsh vegetation will move northwards. Many of the freshwater flora and fauna are more widely distributed than their terrestrial counterparts, and many aquatic plants show considerable ability to survive in various environments. These species therefore are likely to be quite resistant to change.

Some of the most dramatic changes are likely to occur in waterbird populations that breed in the Arctic. An extension of forested landscapes onto the tundra of Alaska, Canada and Siberia – which is predicted by a number of scientists – would reduce available nesting habitat for many species of geese and shorebirds.

Above **During the 1950s, the Aral Sea fishing fleet landed thousands of tonnes of fish each year, now it is just a rusty reminder of the scale of the eco-catastrophe. The shrinking** of the Aral Sea has provided a preview of the likely effects of global climate change. Many wetlands that appear to be secure, in fact exist in a precarious equilibrium.

Creating Awareness

Until people understand why they should safeguard wetland ecosystems and species, as well as being aware of the actions required to do so, many of the changes required to conserve these ecosystems will not take place. The value of the many direct and indirect benefits that wetlands provide, and the social and economic consequences of wetland degradation and loss need, therefore, to be documented and communicated to the widest possible audience.

Priorities and quantification

In the United States, a method for assessing wetlands has been developed to assist in identifying those wetlands that merit the highest conservation priority in the face of urban-industrial developments, such as road construction and house building. While this has proved a useful planning tool, it does not give quantification in economic terms of the functions and value of individual wetlands. Despite the absence of such a methodology, several individuals and institutions have given increasing attention over recent years to the economic quantification of wetlands. A growing number of such studies have been carried out in Europe and North America, and there has been a gradual increase in studies in the tropics.

This work has played a major role among governments and in the development assistance agencies in building initial awareness of the importance of wetland ecosystems, but it still needs to be expanded. To achieve expansion more studies are required, together with an effort to increase the capacity of regional and national training institutions to provide instructions in wetland evaluation.

Work to help address these needs is already under way, notably in Central America, where a manual for the economic assessment of tropical wetlands is in preparation, and in Nigeria, Fiji and Indonesia, where studies of the economic value of specific wetland ecosystems have been undertaken. These initiatives need to be seen as the first steps in what should become a comprehensive programme of action by a diversity of institutions.

The investment in wetland economics just described needs to go beyond the analysis of the value of wetlands to examine the impacts of wetland conservation and management projects. Such analyses are critical if the conservation movement is to demonstrate not only that wetlands are valuable,

Above **Woman milking a cow at a Dinka village near Tonga, Sudan. The Dinka, and the other peoples of the Sudd, follow a traditional lifestyle which has evolved a highly efficient system of wetland agriculture that takes maximum advantage of seasonal floodplains while causing no harm to the environment.**

Right **Women fillet a fresh catch from Lake Turkana, Kenya. Traditional fishing provides a sustainable harvest that has little impact on the lake's fish populations.**

Below **Wetlands provide many human needs. These huts of Kenya's Turkana people have been built from the lake's reeds.**

Left Irrigated corn field in Texas, USA. In many regions of the world, irrigation is essential to the great monocultures that provide food for the bulk of the world's human population. As fresh water becomes an increasingly scarce commodity, stark choices must be made. Water to support natural wetlands diversity; or water to irrigate unnatural, but at present essential, monocultures?

but also that carefully designed investments in wetland conservation can make a significant contribution to local and national economies. Even for environmental economists, such analyses will be pioneering, as only rarely have their studies sought to show who benefits from wetland conservation, and how.

Demonstrating incentives

Ultimately, full understanding of the wise use of wetlands will come about most rapidly through practical demonstrations of their value. In the United States it was only after floodplain destruction resulted in towns flooding, and the dredging and filling of salt marshes reduced fish harvests, that the first state laws regulating wetland alteration were passed. In Zambia, it was the importance of the fishery, pasture and wildlife resources of the Kafue floodplain, and the failure of intensive irrigated agriculture, that led local leaders to argue strongly for the maintenance of the natural floodplain as the most effective way of meeting the needs of the rural people.

On the Kafue flats people opposed wetland destruction because they share in the benefits that the wetlands provide. They did not need a special campaign to convince them of the floodplain's value. Often no amount of effort to build awareness of wetlands in the abstract will change people's behaviour, rather they must actually see the benefits of well-preserved wetlands. Conservation, therefore, needs to be built around these benefits, designing and implementing economic incentives in conjunction with awareness campaigns that publicize these incentives.

Communication of information

Information generated about wetlands will be of little value if it is not communicated effectively to a diversity of audiences. A carefully thought-through communications strategy is required for wetland conservation. Understanding of wetlands needs to be built at all levels of society. Three audiences merit particular attention: the general public; local communities dependent upon wetland resources; and the government departments and non-governmental agencies that make decisions on investment in wetland conservation and development.

The widest possible range of conservation organizations need to use their communicative strengths to improve the flow of information on wetlands and target this at critical audiences. This can be achieved through the production of specific materials on wetlands, their value and the public cost of wetland loss, and by dissemination of these and other materials to the different audiences. The materials will need to explain both the importance of wetlands and the relevance of wetland conservation to critical environment and development issues. These issues include wetlands and demographic change, wetlands and freshwater management, tidal wetlands and sea-level rise, and wetland biodiversity.

Building National Capacity

One of the most significant changes in the field of wetland conservation in recent years has been the increasingly interdisciplinary approach to management. Much of the concern for wetland conservation has its origin in the importance of wildlife, and wetland conservation was for decades the exclusive domain of biologists and protected-area managers. With growing awareness of the broader value of wetlands to human society, however, a much wider range of institutions, including those concerned with fisheries, agriculture and livestock, have become engaged in wetland conservation. At the same time, recognition that wetlands need to be maintained as functioning units in the landscape – rather than simply in a limited number of national parks and other protected areas – has led in many countries to a growing understanding that conventional approaches to wetland conservation are largely inadequate.

The appreciation that new partnerships and new approaches are required is central to many of the most significant recent achievements in wetland conservation. For example, the North American Waterfowl Management Plan is based upon an alliance of traditional conservation groups, farming and industry; similarly, in Uganda, the National Wetlands Programme has brought together an alliance of organizations involved with fisheries, forestry, water resources, agriculture, national parks and other environmental issues.

Wetland programmes

A growing number of countries have begun the process of building national wetland programmes. The central element in this process has been consultation at all levels, including a diversity of government agencies, non-

Left **Fish ladder at the side of the Bonneville Dam, on the Columbia River, Oregon, USA. The needs of migrating trout and salmon seem insignificant compared with the thousands of megawatts of electricity, lighting millions of homes and factories, produced by the Columbia dams. The trout and salmon are, however, an essential part of several important food chains. Among the top predators that feed on the fish are grizzly bears, bald eagles and human beings.**

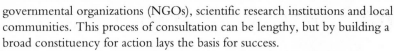

Left **Wetlands are naturally resilient, and often contain the seeds of their own survival. Along the banks of the River Tees in northern England, banks of reeds have been planted in order to help detoxify the riverside marshes.**

Below **In Botswana's Okavango Delta region, hippo-proof fences are used to restrict the animals' movement. The hippos are a major vector in the spread of the choking aquatic weed *Salvinia molesta*.**

The concern for wetland conservation has its origins in the importance of wildlife, . . . and was for decades the exclusive domain of biologists and protected-area managers.

governmental organizations (NGOs), scientific research institutions and local communities. This process of consultation can be lengthy, but by building a broad constituency for action lays the basis for success.

The most important step, of course, is to get the process under way. In some countries senior ministerial concern has been the driving force for this, leading to a national agency taking over coordinating responsibility. In other countries, NGOs have encouraged their governments to take steps by convening national wetland workshops, while in others consultative processes have from the outset been joint efforts by both government and non-governmental groups. The critical message of this experience is that while effective consultation is central to success, this can be achieved in a number of ways. The challenge for the next 10 years will be to build upon these lessons and initiate similar processes in a much greater number of countries. We must also ensure that both new and continuing schemes are effectively supported. This will require substantial investment and a concerted effort by international institutions.

The role of NGOs

Non-governmental organizations have provided a central force for wetland conservation in many countries and regions. For example, the Kenya Wetland Working Group has played a leading role in encouraging the development of a national wetlands programme; in the Americas, the Western Hemisphere Shorebird Reserve Network has provided an exceptionally effective forum for international collaboration on coastal wetland conservation. The role of NGOs is already significant, but it is likely that in the climate following the United Nations Conference on Environment and Development, and with pressures upon wetlands continuing to grow by the day, the importance of the independent, and often politically unpopular, voice of these organizations will continue to grow. Special attention therefore needs to be given to providing the NGOs with the necessary technical, logistical and financial support.

Institutional capacity

Even where awareness of the importance of wetlands is well developed among senior decision-makers, it often cannot be translated into effective conservation action because of the limited institutional capacity to conserve and manage wetlands. Little long-term progress can be achieved without sustained investment in building the institutional capacity to manage wetland resources.

There are two principal ways in which institutions need strengthening. First, institutional arrangements that support an integrated approach to wetland management need to be developed. In some countries, such as Uganda, this has involved the establishment of inter-ministerial wetland committees that provide a forum for debate between different ministries and departments with wetland responsibilities. While these mechanisms have proved successful, they are essentially voluntary mechanisms for collaboration. The use of supporting legislation, therefore, needs to be explored to determine where it can assist in the functioning of such mechanisms.

The second focus of effort to strengthen institutional capacity is through training and staff development. In many countries this is the form of international support of wetland management that is sought most frequently, and many international institutions invest in wetland training each year. Only rarely, however, have these efforts formed part of a programme targeted at the most important institutional weaknesses; neither have they been followed up with other forms of institutional support, such as funding and political support. Steps are being taken to address these deficiencies, and a growing number of national and regional training activities are being pursued. Much greater investment is required, however, if this is to achieve a lasting impact that will help to protect the countless thousands of endangered wetlands.

International Cooperation

Over the course of the next 25 years, many of the initiatives begun in recent years will continue and expand, driven by national concern for wetland conservation. In most cases this action will be enhanced in its effectiveness by the exchange of information between the countries and institutions concerned with these issues, either in the same or in different regions. In many cases close collaboration will be essential to address the problems identified. In some regions, mechanisms already exist for collaboration and function well. The lessons from these need to be assimilated and promoted.

In other regions however, cooperative mechanisms are at best weak and are often absent. In West Africa, disagreements over the Senegal floodplain led to armed conflict, and in many countries freshwater management remains a source of considerable tension. These and other problems indicate that while international cooperation is essential, considerable flexibility needs to be exercised in identifying priorities for investment and in choosing the mechanisms to address these.

The Ramsar Convention

In the context of international cooperation, the Ramsar Convention has a central role to play, providing the single most important framework for intergovernmental cooperation on wetland issues. The cooperation between the Netherlands, Germany and Denmark over the Wadden Sea, and between Canada, the United States and Mexico over the North American Waterfowl Management Plan have already made use of this framework and emphasized the obligations for cooperation between contracting parties. Other opportunities for cooperation under the Ramsar framework are being explored. These will need to complement similar efforts being carried out under other intergovernmental frameworks, such as the Environmentally Sound Management of Inland Water (EMINWA) programme of the United Nations Environment Programme (UNEP) on the integrated management of inland waters, notably large lakes and river systems. Similarly, regional institutions such as the Southern African Development Coordination Conference (SADCC) and the Association of Southeast Asian Nations (ASEAN) have their own regional wetland programmes – with support from IUCN and Australia respectively. In regions where there are few contracting parties to the Ramsar Convention, this kind of approach will, for the immediate future, provide the most effective mechanisms for regional cooperation on wetland issues.

Development assistance policies

An important area that has so far received limited attention under the framework of the Ramsar Convention is development assistance and its impact upon wetland ecosystems. The United States Treasury Department has issued "voting standards", which instruct the United States Executive Directors to the multilateral banks and the Administrator of the United States Agency for International Development (USAID) not to provide support to projects that will harm wetlands and to encourage those that will help conserve wetlands. However, the United States is so far the only contracting party to have established an

Left **The Bijagos Archipelago is being considered as an area for tourist development. With many countries attracting foreign development investment, careful planning of tourism is desperately needed.**

Bottom left **Swamps support some of Africa's most productive fisheries. Communities like this village in Benin would benefit from an approach that had encouraged small-scale aquaculture as well as natural fisheries.**

Below **Pine forest marshland in Finland. With the highest proportion of wetlands in Europe, Finland is at the fore of wetland conservation, and lends its support to many projects all over the developing world.**

explicit aid policy in support of its obligations under the Ramsar Convention. Moreover, the impact of this policy has been limited so far, and mechanisms to allow its widespread application are lacking.

Although no European Contracting Party has yet produced an explicit aid policy for wetlands similar to that of the United States, several aid agencies currently provide substantial support to wetland conservation in the developing world. The Netherlands, Norway, Finland and Switzerland are notable in this area and provide a model for other countries. Such actions are encouraging, but they are likely to have only a limited effect on wetland loss worldwide unless accompanied by policies limiting investment in projects that destroy and degrade wetlands. Expenditure of US$10 million per year in wetland conservation will have little long-term impact in the face of several hundred million dollars being spent on dams, polders, aquaculture and other projects that destroy wetlands and marginalize the rural communities that depend upon them.

The integrated approach

The Ramsar Convention provides the key international agreement under which to promote greater investment in wetland management through development assistance programmes. However, technical support from IUCN, the International Waterfowl and Wetlands Bureau and other organizations with expertise in the integration of conservation is essential, especially to convince development professionals that alternative approaches to wetland management can yield economic and social benefits.

At the same time, the capacity of national institutions to seek and use increased funding for wetland conservation needs to be increased. This requires expanded support for field projects and training programmes that can serve as the catalysts to drive this process. Perhaps most important of all is the promotion of the results of field projects at an international level; this should form part of a substantially increased effort to convince decision makers in government and the development assistance community that wetland conservation yields substantial results for social and economic development.

Index

Species index

Sources

Baldock, D. 1990
Agriculture and Habitat Loss in Europe.
WWF/EEP. 60pp.

Collins, M. 1990
The Last Rain Forests. Mitchell Beazley

Dugan, P.J. 1990
Wetland Conservation: a review of current issues
and required action.
IUCN, Gland, Switzerland. 96pp.

Finlayson, M. & M.E. Moser. 1991
Wetlands. Facts on File, UK

Hoskin, J. & A.W. Hopkins. 1991
The Mekong: a river and its people.
Post Publishing, Bangkok

Howell, P.M. Lock & S. Cobb (Eds). 1988
The Jonglei Canal: impact and opportunity.
Cambridge University Press

Hughes, R.H. & J.S. Hughes. 1992
A Directory of African Wetlands.
IUCN, Gland, Switzerland and Cambridge,
UK/UNEP, Nairobi, Kenya/WCMC Cambridge,
UK, xxxxiv + 820pp, 48 maps

Jamsai, S. 1988
Naga: cultural origins in Siam and the West Pacific

Lee, D.B. 1990
Okavango Delta: old Africa's last refuge.
Nat. Geog. Dec. 38-69

Löfroth, M. 1991
Våtmarkerna och deras betydelse (English summary:
Wetlands and their importance) -
Swedish EPA Report 3824, 91pp

McComb, A.J. & P.S. Lake. 1988
The Conservation of Australian Wetlands.
Sarry Beatty & Sons Ltd, Australia

Moss, B. 1980
Ecology of Freshwaters.
Blackwell, Oxford. 332pp

Pearce, F. 1991
The rivers that won't be tamed.
New Scientist, 13 April pp 38-41

Pearce, F. 1991
Acts of God, acts of man?
New Scientist, 18 May pp 20-21

Ross, K. 1987
Okavango: Jewel of the Kalahari.
BBC Books, London. 256pp

Schaller, G.B. 1983
Mammals and their biomass on a Brazilian Ranch.
Arquivos de Zoologia, Sao Paulo, 31 910; 1-36

Schaller, G.B. 7 J.M.C. Vasconcelos. 1978
A Marsh Deer Census in Brazil.
Vol. 14: 345-351

Scott, D.A. & M. Carbonnell. 1986
Neotropical Wetlands Directory, IWRB.
Slimbridge and IUCN, Cambridge

Scott, D.A. 1989
A Directory of Asian Wetlands.
IUCN, Gland

Struzik, E. 1992
The rise and fall of Wood Buffalo National Park.
Borealis 10: 11-25

Thorbjarnarson, J. 1992
Crododiles: An Action Plan for their Conservation.
IUCN, Gland, Switzerland

1991
The Environment. National Atlas of Sweden.
Stockholm, 184pp

1991
Acidification and liming of Swedish freshwaters.
Monitor 12, Swedish EPA Informs, 144pp

1991
Lake Hornborga - past, present and future -
Swedish Informs, 8pp

Artworks

p11, p16-17 Colin Rose; p18 Abigail Edgar;
p41 Malcolm Mc Gregor;
p54-55, p98, p73, p102, p107
Jean Jottrand and Caroline Wollen

Map Acknowledgements

Wetlands in Danger has required the input of a
number of staff at WCMC, in particular, Ian Barnes,
Clare Billington, Simon Blyth, Gilian Bunting,
Mary Edwards, Corina Ravilious and Mark
Spalding. The project was devised by Mark Collins
and managed by Richard Luxmoore. Advice on map
contents was sought from Max Findlayson, Geoff
Howard, Bob Hughes, Ed Maltby and Derek Scott.

Data on the important bird areas were supplied by
the International Council for Bird Preservation and
for the Ramsar Sites by the International Wildfowl
and Wetland Research Bureau. A digitized map
of the wetlands of Mexico was developed by
Conservation International and the Secretaria de
Desarrollo Urbano y Ecologia with the funding
from the United States Fish and Wildlife Service.

Collection of data for the maps was made possible
by sponsorship from the British Petroleum to enable
the development of the WCMC Biodiversity Map
Library.

Picture Credits

(Abbreviations: R=Right, L=Left, T=Top, C=Centre,
B=Bottom/Below)

1 Oxford Scientific Films/Carol Farnett/Partridge
Films; 2-3 Planet Earth Pictures/John Downer;
6-7 Oxford Scientific Films/M.Wendler/Okapia;
8-9 Oxford Scientific Films/Carol Farnett; 12 Planet
Earth Pictures/Chris Huxley; 13T Robert Harding
Picture Library; 13B D.A.Scott; 14-15B Robert
Harding Picture Library; 14-15T Luonnonkuva-
Arkisto; 15C Planet Earth Pictures/David George;
16BL Planet Earth Pictures/Peter Scoones; 17TR
Planet Earth Pictures/John Lythgoe; 18-19B Oxford
Scientific Films/Philippe Henry; 19 Planet Earth
Pictures/Robert Cairns; 20-21 Robert Harding
Picture Library; 22-23 Frank Spooner Pictures;
24TL Planet Earth Pictures/Brian Alker; 24-25T
Zul, Chapel Studios; 26T Planet Earth Pictures/
Brian Alker; 26-27B Planet Earth Pictures/John
Lythgoe; 27CL Oxford Scientific Films/Tony
Bomford & Tim Borrill; 28-29 Planet Earth
Pictures/Colin Pennycuick; 30-31B Luonnonkuva-
Arkisto; 30 Planet Earth Pictures/ Mark Mattock;
31T Planet Earth Pictures/David A. Ponton;
32T Robert Harding Picture Library; 32-33B
Planet Earth Pictures/Robert Canis; 33CL Planet
Earth Pictures/Mark Mattock; 34-35B Planet
Earth Pictures/Robert Canis; 34CR Planet Earth
Pictures/Mark Mattock; 35CL Planet Earth Pictures/
P.J.Palmer; 36B Planet Earth Pictures/ Franz J.
Camenzind; 36CR Planet Earth Pictures/ Philip
Chapman; 37 Planet Earth Pictures/Doug Perrine;
38BL Planet Earth Pictures/Jonathan Scott Earth;
38-39B Planet Earth Pictures/Pete Atkinson;
38TR Zul, Chapel Studios; 40T Planet Earth
Pictures/John Lythgoe; 41B Survival Anglia/Alan
Root; 42-43 Oxford Scientific Films/C.C.
Lockwood/Earth Sciences; 44-45B IUCN; 44CR
Oxford Scientific Films/Jack Dermid; 45TR Oxford
Scientific Films/John McCammon; 46-47 Oxford
Scientific Films/Richard Packwood; 46-47B
ZEFA; 47BR IUCN; 48TL IUCN; 48-49T Oxford
Scientific Films/Stouffer Productions Ltd; 49B
ZEFA; 50-51B Robert Harding Picture Library;
50-51T Oxford Scientific Films/Harold Taylor;
51CR Planet Earth Pictures/John Downer; 52-53
NHPA/ Stephen Krasemann; 57 ZEFA; 60-61T
Planet Earth Pictures/Nigel Tucker; 60-61B ZEFA;
61BR Oxford Scientific Films/Dan Guravich; 64BL
ZEFA; 64-65 ZEFA; 65TR Robert Harding Picture
Library; 66-67 N.A.W.C.C.(Canada); 66-67T
Wilfried Schurig; 66-67B Wilfried Schurig; 67T
Oxford Scientific Films/Bates Littlehales/Earth
Sciences; 67TR Wilfried Schurig; 70-71T Frank
Spooner Pictures; 70-71B ZEFA; 72BR Oxford
Scientific Films/Earth Sciences; 72TR Oxford
Scientific Films/Tom Ulrich; 74-75B Oxford
Scientific Films/C.C. Lockwood, Earth Sciences;
74BL Planet Earth Pictures/James D.Watt; 74-75T
Planet Earth Pictures/John Lythgoe; 78-79B IUCN;

78T IUCN; 79TR IUCN; 80-81T IUCN; 80-81B
IUCN; 81BR Planet Earth Pictures/Linda Pitkin;
81CR Planet Earth Pictures/Linda Pitkin; 82BR
Planet Earth Pictures/Ken Lucas; 82TL Frank
Spooner Pictures; 82-83C IUCN; 86B Oxford
Scientific Films/Aldo Branso Leon; 87C Planet Earth
Pictures/Andre Bartschi; 87B Planet Earth Pictures/
Paulo Oliveira; 86T Planet Earth Pictures/ Andre
Bartschi; 88-89B South American Pictures/ Tony
Morrison; 88-89T Planet Earth Pictures/Ken Lucas;
90-91T Planet Earth Pictures/Richard Matthews;
90BR ZEFA; 91B ZEFA; 94-95 Robert Harding
Picture Library; 95T Robert Harding Picture
Library; 95B Planet Earth Pictures/Chris Huxley;
96-97B Oxford Scientific Films/M. Wendler;
96-97T Planet Earth Pictures/Richard Matthews;
97TR Zul, Chapel Studios; 98-99B Oxford
Scientific Films/Ronald Templeton; 99 Oxford
Scientific Films/Partridge Productionc; 102-103T
Luonnonkuva-Arkisto; 103B Jan Johansson; 106BL
Robert Harding Picture Library; 106-107 ZEFA;
108C Benelux Press BV; 108-109T Benelux Press
BV; 109B Benelux Press BV; 110-111B Explorer/
Hervy; 110BL Robert Harding Picture Library;
111TL Planet Earth Pictures/Jonathan Scott;
112-113T Oxford Scientific Films/Richard
Packwood; 112BL Francoise Sarano, Equipe
Cousteau; 113BL Robert Harding Picture Library;
113 TR Oxford Scientific Films/Stephen Dalton;
114C Robert Harding Picture Library; 114-115B
Planet Earth Pictures/Geoff du Feu; 115T Planet
Earth Pictures/Nigel Downer; 118TR IUCN;
118-119B Robert Harding Picture Library;
119TL Robert Harding Picture Library; 120-121B
Explorer/A.Audano; 120T Robert Harding Picture
Library; 121T Oxford Scientific Films/Michael
Fogden; 124TL ZEFA; 124-125B ZEFA; 125TR
Robert Harding Picture Library; 126-127C
A.S.A.P./Garo Nalbandian; 126-127B Robert
Harding Picture Library; 127T Oxford Scientific
Films/David Cayless; 130TR Planet Earth Pictures/
Sean Avery; 130-131 Planet Earth Pictures/John
Downer; 131TL ZEFA; 134TR Robert Harding
Picture Library; 134-135B Robert Harding Picture
Library; 135TL ZEFA; 136-137B IUCN; 136TL
IUCN; 136-137T IUCN; 138-139B IUCN;
138-139T ZEFA; 139BR ZEFA; 142TL IUCN;
142-143B Planet Earth Pictures/Nick Greaves;
142TR Planet Earth Pictures/Keith Scholey;
144-145B Planet Earth Pictures/Jeannie
Mackinnon; 144-145T Planet Earth Pictures/J.R.
Bracegirdle; 145BL Planet Earth Pictures/Richard
Coomber; 148-149T Colorific/David Turnley,
Black Star; 148TL Frank Spooner Pictures;
148-149B ZEFA; 152T IUCN; 152-153 Robert
Harding Picture Library; 153TL Planet Earth
Pictures/Anup Shah; 154-155 Robert Harding
Picture Library; 154BR Robert Harding Picture
Library; 155 Robert Harding Picture Library;
158-159T Robert Harding Picture Library;
158-159B Robert Harding Picture Library;
159TR Oxford Scientific Films/Michael Fogden;
160 Planet Earth Pictures/Sue Earle; 161T Oxford
Scientific Films/Frank Schneidermeyer; 161B Planet
Earth Pictures/Sue Earle; 164 Magnum/c.Hiroji
Kubota; 165B Network/Klaus Bossemeyer/
Bilderberg; 165TL Zul, Chapel Studios; 166CR
Oxford Scientific Films/Animals Animals/Zig
Lesczynski; 166-167T Planet Earth Pictures/Keith
Scholey; 167B Still Pictures/Bojan Brecilj; 170TR
Planet Earth Pictures/Hans Christian Heap;
170-171B Wildlight/Grenville Turner; 171CL
Oxford Scientific Films/Roger Brown; 174-175B
Robert Harding Picture Library; 174-175T Zul,
Chapel Studios; 176-177 Explorer/Francois
Jourdan; 178-179B Explorer/Francois Jourdan;
179T ZEFA; 180CR Planet Earth Pictures/Mary
Clay; 180-181T Rex Features; 181B W.W.F.
Photolibrary/John E. Newby; 182TR Robert
Harding Picture Library; 182-183B ZEFA; 183CL
ZEFA; 183TL ZEFA; 184-185T Oxford Scientific
Films/Mike Birkhead; 184-185B Oxford Scientific
Films/Karen Ross; 184BL Oxford Scientific Films/
Breck P. Kent; 186-187T IUCN; 186-187B ZEFA;
187TR Luonnonkuva-Arkisto

Every effort has been made to trace the copyright
holders and we apologise in advance for any
unintentional omissions. We would be pleased to
insert the appropriate acknowledgement in any
subsequent edition of this publication.